Martin Slowik

Metastability in Stochastic Dynamics

Martin Slowik

Metastability in Stochastic Dynamics

Contributions to the Potential Theoretic Approach

Südwestdeutscher Verlag für Hochschulschriften

Impressum / Imprint
Bibliografische Information der Deutschen Nationalbibliothek: Die Deutsche Nationalbibliothek verzeichnet diese Publikation in der Deutschen Nationalbibliografie; detaillierte bibliografische Daten sind im Internet über http://dnb.d-nb.de abrufbar.
Alle in diesem Buch genannten Marken und Produktnamen unterliegen warenzeichen-, marken- oder patentrechtlichem Schutz bzw. sind Warenzeichen oder eingetragene Warenzeichen der jeweiligen Inhaber. Die Wiedergabe von Marken, Produktnamen, Gebrauchsnamen, Handelsnamen, Warenbezeichnungen u.s.w. in diesem Werk berechtigt auch ohne besondere Kennzeichnung nicht zu der Annahme, dass solche Namen im Sinne der Warenzeichen- und Markenschutzgesetzgebung als frei zu betrachten wären und daher von jedermann benutzt werden dürften.

Bibliographic information published by the Deutsche Nationalbibliothek: The Deutsche Nationalbibliothek lists this publication in the Deutsche Nationalbibliografie; detailed bibliographic data are available in the Internet at http://dnb.d-nb.de.
Any brand names and product names mentioned in this book are subject to trademark, brand or patent protection and are trademarks or registered trademarks of their respective holders. The use of brand names, product names, common names, trade names, product descriptions etc. even without a particular marking in this works is in no way to be construed to mean that such names may be regarded as unrestricted in respect of trademark and brand protection legislation and could thus be used by anyone.

Coverbild / Cover image: www.ingimage.com

Verlag / Publisher:
Südwestdeutscher Verlag für Hochschulschriften
ist ein Imprint der / is a trademark of
AV Akademikerverlag GmbH & Co. KG
Heinrich-Böcking-Str. 6-8, 66121 Saarbrücken, Deutschland / Germany
Email: info@svh-verlag.de

Herstellung: siehe letzte Seite /
Printed at: see last page
ISBN: 978-3-8381-3412-3

Zugl. / Approved by: Berlin, TU, Diss., 2012

Copyright © 2012 AV Akademikerverlag GmbH & Co. KG
Alle Rechte vorbehalten. / All rights reserved. Saarbrücken 2012

Contents

Preface   iii
   Motivation   iii
   An outline of this book   ix
   Acknowledgements   xii

Chapter 1. Basic ingredients of the potential theoretic approach   1
   1.1. Introduction   2
   1.2. Boundary value problems   5
   1.3. Laplace equation and Poisson equation   10
   1.4. Green function and Green's identities   12
   1.5. Equilibrium potential, capacity, and mean hitting times   14
   1.6. The Dirichlet and Thomson principle: a dual variational principle   18
   1.7. The Berman-Konsowa principle   21
   1.8. Renewal equation   23
   1.9. Almost ultrametricity of capacities   27

Chapter 2. Coupling arguments and pointwise estimates   31
   2.1. Introduction   32
   2.2. Construction of the Coupling   35
   2.3. Bounds on harmonic functions and local recurrence   42
   2.4. Cycle decomposition of σ-paths   43
   2.5. Bounds on the mean hitting time and the Laplace transform   48

Chapter 3. Metastability		**51**
3.1.	Introduction	52
3.2.	Ultrametricity	55
3.3.	Sharp estimates on mean hitting times	58
3.4.	Asymptotic exponential distribution	62
Chapter 4. Metastability and Spectral Theory		**69**
4.1.	Introduction	70
4.2.	Rough localization of eigenvalues	76
4.3.	Characterization of small eigenvalues	82
Chapter 5. Random field Curie–Weiss–Potts model		**89**
5.1.	Introduction	90
5.2.	Induced measure and free energy landscape	96
5.3.	Upper bounds on capacities	108
5.4.	Lower bounds on capacities	123
5.5.	Sharp estimates on metastable exit times and exponential distribution	148
Bibliography		**151**

Preface

This book is concerned with several aspects of metastability in stochastic dynamics with a particular focus on the dynamical behavior of disordered mean field spin systems at finite temperature.

Motivation

Phenomenology. *Metastable behavior* of a complex system, either natural or artificial, is causal for a number of interesting phenomena as diverse as delayed freezing of super cooled liquids, folding of large bio-molecules, kinetics of chemical reactions, changes in global climate systems, apparent stability of stock markets, to name but a few. Despite such a diversity of scientific areas ranging from physics, chemistry, biology to economics in which metastable behavior can be observed, the common feature of all these situations is the existence of multiple, well separated time scales. On the corresponding short time scales, the system can only explore at limited part of the available state space. Depending on the time scale under consideration, the state space can be decomposed into several disjoint regions, in which the system is effectively trapped. Within such a region, that may be viewed as *metastable set* to which one may associate a *metastable state*, the system appears to be in a *quasi-equilibrium*. At larger time scales, rapid transitions between metastable states occur which are induced by *random fluctuations*. The main mathematical task we want to address is to analyse such a system on long time scales. In particular, we need some understanding of how the process manages to escape from a metastable state.

In order to illustrate the phenomenon of metastability in more detail, let us consider the *dynamical* behavior of a ferromagnet close to its *first order phase transition*, see [97, 94]. In the framework of equilibrium statistical mechanics, the defining characteristic of a first-order phase transition is a discontinuity in an extensive variable, such as the magnetization, as a function of an intensive variable, for instance the magnetic field.

It is well known that for a ferromagnet above the critical temperature, T_c, there exists an unique *paramagnetic* phase. Experimentally, one observes that the magnetization $m(h)$ vanishes when switching off the external magnetic field h, i.e. $m(h) \to 0$ as $h \to 0$. Below T_c the behavior of the system is markedly different. Depending on whether the external field approaches zero from below or from above there exist two different phases that can be distinguished by their different magnetization,

$$\lim_{h \downarrow 0} m(h) = m_+ > 0, \qquad \lim_{h \uparrow 0} m(h) = m_- < 0.$$

This phenomenon is called *spontaneous* magnetization. Now, consider the following experiment. Suppose the system is prepared in such a way that it reveals the magnetization $m(0) = m_+$. After switching on a small negative magnetic field $|h| < h_c$, one may observe that instead of undergoing a rapid phase transition the system remains in an apparently stationary state for a (macroscopic) long time that is characterized by a positive magnetization $m(h) > 0$. This situation corresponds to a metastable state. In contrast to that, the magnetization of the equilibrium state has the same sign as the external field $h < 0$.

In order to avoid the impression that a metastable behavior can be described only by considering static aspects, e.g. an analytic continuation of the critical curves in the phase diagram, let us stress the fact that metastability is a *dynamical* phenomenon.

For this reason, let us now describe the dynamical behavior of a magnetic system in a metastable state. On a microscopic level, the magnetization of an individual (classical) spin can change its sign from $+1$ to -1 and vice versa induced by internal thermal perturbations. For this reason, at any time small droplets of downward pointing spins appear in a "sea of upward-pointing spins". While the spins in the bulk of such a droplet become aligned with the external field, spins at the surface feel both the negative external field and an positive field coming from spins outside the droplet. In other words, there is a competition between the gain of energy in the bulk of a droplet and the loss of energy at its surface. This causes the dissolution of small droplets whereas a large droplet, once created, can be seen as a gate through which the system escapes rapidly from the metastable to the stable equilibrium state. However, the spontaneous formation of a sufficiently large droplet is a rather rare event which may be seen as an explanation that an escape from a metastable state can only be observed on macroscopic time scales.

Metastability and diffusions. From a mathematical point of view, "Kramers' equation has become the paradigm of metastability" [1]. For this reason, we describe in the sequel briefly some aspects of its derivation and some of the rigorous mathematical tools that have been invented in order to study it. Due to the interdependence between experimental observations, rigorous/non-rigorous theoretical analysis and computer simulations of metastable systems, it is rather challenging to give proper credit to the people involved. In particular, the review presented below is far from being complete. For more a detailed review and an outlook of future challenges, we refer to the recent papers [12, 99].

The origin of this equation can be traced back to the study of kinetic of chemical reactions. In most real-world applications, metastable systems are described by many-particle systems. Since the dynamics of a many-particle system is very difficult to analyse either analytically or numerically, model reductions play an important role. For instance, the full phase space of a system describing the dynamics of a chemical reaction has $\sim 10^{23}$ degrees of freedom. Therefore, instead of studying the dynamics on the high-dimensional surface in the full phase space Σ, one of the first model reduction steps is to consider the dynamics only along the so-called *reaction path* $\chi = (X_t, \dot{X}_t) \in \Sigma$ in an effective potential that takes into account the interaction with the (thermal) reservoir. As a consequence of such a projection from Σ to χ, the resulting dynamics is in general described by a non-Markovian process [57, 58]. Assuming additionally that the noise correlation times in the reservoir are extremely short, one can further use a Markovian approximation for the reduced dynamics [116].

These reduction steps lead to the *first mathematical model* for metastability proposed in 1940 by Kramers [78] in order to describe a chemical reaction. It consists of a classical particle of mass one moving in an one-dimensional asymmetric double-well potential U under the influence of Gaussian white noise and friction with coefficient γ which models effectively the thermal reservoir at temperature T. Its equation of motion is given by

$$\gamma^{-1}\ddot{X}_t = -\dot{X}_t - U'(X_t) + \sqrt{2\varepsilon}\,\dot{B}_t,$$

where the parameter $\varepsilon \equiv \varepsilon(T)$ is temperature dependent and tends to zero when T approaches the absolute zero. In the limit when the friction becomes infinitely strong, this equation results in the simple, one-dimensional diffusion equation

$$dX_t = b(X_t)\,dt + \sqrt{2\varepsilon}\,dB_t \qquad (0.1)$$

to which we refer to as *Kramers' equation*, where $b(x) = -U'(x)$ and B_t is assumed to be a Brownian motion. Notice that local minima of U correspond to metastable

---

[1] Bovier [12, p.2]

states/points in this model. Kramers studied various interesting questions in the context of this model. In particular, he identified the mechanism of noise-assisted reactions and derived an explicit formula, also called *Kramers formula*, for the averaged transition time from the local minimum a to the global minimum b passing through the maximum z that lies in between them,

$$\mathbb{E}_a[\tau_b] = \frac{2\pi}{\sqrt{U''(a)|U''(z)|}} \exp\left(\varepsilon^{-1}\left(U(z) - U(a)\right)\right)(1 + o_\varepsilon(1)). \qquad (0.2)$$

Based on experimental data, Arrhenius had already suggested in 1889 that the reaction-rates are proportional to the inverse of the temperature on logarithmic scales [2]. The exponential term in (0.2) reflects *Arrhenius' law*. Let us now have a look at the prefactor. In comparison with a sharply-peaked maximum, it is more likely that the particle, when slightly passing by a flat maximum, returns to its starting well. On the other hand, in a flat minimum the particle gets less often close to the barrier than in a peaked one. This explains why a smaller curvature of the local minimum as well as the maximum leads to a larger averaged transition time.

Let us remark that, in the context of *reaction rate theory*, Eyring had given previously a heuristic derivation of the prefactor based on quantum and classical statistical mechanics computations expressing the prefactor as the ratio of the partition function of the reactants and the "activated complex", respectively, [47]. For a discussion of past and possible future developments in the reaction rate theory we refer the reader to the review by Hänngi, Talkner and Borkovec [61] and the recent paper by Pollak and Talkner [99].

In the context of randomly perturbed dynamical systems, generalizing Kramers' equation to higher dimensions, various dynamical aspects concerning the exit problem of a domain were first analysed rigorously in the seminal work by Freidlin and Wentzell [111, 51]. They invented the idea to use *large deviations in path space* as well as to consider a *Markov chain with exponential small transitions* to model effectively the jumps between different attractors. Let us denote by Γ the set of all path, $\gamma : [0, T] \to \mathbb{R}^d$ with arbitrary T. Employing large deviation principles in path space, one can control the probability that the process, $\{X_t\}$, stays close to a given path $\gamma \in \Gamma$ over a time interval $[0, T]$, in the sense that

$$\varepsilon \ln \mathbb{P}\left[\sup_{t \in [0,T]} \|X_t - \gamma_t\| \leq \delta\right] = -I(\gamma) + o_\varepsilon(1), \qquad (0.3)$$

for $\delta > 0$ and $I : \Gamma \to \mathbb{R}_+$ a lower semi-continuous function with compact level sets. For studying a transition between the disjoint neighbourhood A, B of two different local minima, the task is to compute the optimizer of the variational problem $\inf_{\gamma: A \to B} I(\gamma)$ and to analyze its properties. Notice the analogy, at least to some extend, between this approach and the reaction path considered previously. Various interesting aspects of metastability, e.g. properties of the typical exit path and the asymptotic exponential distribution of exit times, was later proven for random perturbations of dynamical systems of Freidlin-Wentzell type in finite dimensions [52], infinite dimensions [24, 48, 20] and on Riemannian manifolds [107] by means of large deviations in path space. This method has been proven to be robust and rather universal applicable in many different model contexts [77, 92, 93]. For an in-depth overview we refer to a recent monograph by Olivieri and Vares [94].

A limitation of large deviation methods is its precision. It allows to compute for instance mean exit times only up to an multiplicative error of order $\exp(o_\varepsilon(1))$. For this reason, it is not possible to resolve the behavior of the process near the saddle point. However, this is an essential task in order to derive rigorously the prefactor in the *Eyring-Kramers formula*, the higher-dimensional analog of (0.2). An alternative to the method of large deviations in path space, yielding more precise results beyond logarithmic equivalence, is the *potential theoretic approach to metastability*, systematically developed by Bovier and co-authors, see [10, 11] for a detailed introduction. Its key idea is to express probabilistic quantities of interest in terms of *capacities* and use variational principles to compute the latter. In particular, for reversible diffusion processes of Freidlin-Wentzell type, Bovier, Eckhoff, Gayrard and Klein demonstrated in [17] that sharp estimates of capacities can be easily obtained. As a consequence, they gave a first rigorous proof of the prefactor in the classical Eyring-Kramers formula for dimension $d \geq 1$ and sharp asymptotic estimates of the exponentially small eigenvalues. Later, Berglund and Gentz relaxed the assumptions on the relevant saddle point allowing certain non-degeneracies [56].

Recently, Helffer, Nier and Klein have developed a new analytic approach based on *hypoelliptic techniques* initially developed for the regularity analysis of partial differential equations [63]. In the reversible diffusion setting and under suitable assumptions on the potential U, they derived rigorously a complete asymptotic expansion of the mean hitting times in powers of ε using the so-called *Witten complex* [62].

Metastability and models from statistical mechanics. In the sequel, we will give some more points of the enormous work on metastability that has been done in the past 40 years. In contrast to the diffusion setting, where metastable states corresponds to local minima of the potential U, in models coming from statistical mechanics the question of how to characterize metastable states, and how to identify them in a given model context is really an issue.

A first rigorous formulation of metastable states, taking the dynamical aspect into account, dates back to the work of Penrose and Lebowitz in 1971 on the van der Waals limit of Kac potentials [96, 97]. Subsequently, early results on the metastable behavior of the two-dimensional Ising model were obtained by Capocaccia, Cassandro and Olivieri [22].

In [23] Cassandro, Galves, Olivieri and Vares proposed a different approach to metastability that is based on the pathwise analysis of the dynamics. As a first examples, they demonstrated the efficiency of this concept in the Curie-Weiss model describing its dynamical properties. In the sequel, this approach was successfully applied to more realistic models. In the pioneering works [89, 90], Neves and Schonmann studied the metastable behavior of the two-dimensional Ising model under Glauber dynamics in finite volume and very low temperature. This work was later extended to higher dimensions [3, 25], infinite volume but low temperature [32, 33], infinite volume and fixed temperature but vanishing external field [105] and probabilistic cellular automata [28]. In contrast to Glauber dynamics, Kawasaki dynamics is *conservative*, i.e. the total number of particles is fixed. In a series of papers den Hollander et al. [35, 34] and Gaudillière et al. [54] investigated in the nucleation and metastability for particle systems under Kawasaki dynamics in a large box in two and three dimensions at low temperature and low density.

A more recent approach to metastability was developed by Bovier, Eckhoff, Gayrard and Klein [15, 16]. Based on techniques from potential theory, this approach allows to compute sharp estimates for metastable exit times and their asymptotic exponential distribution. Moreover, it establishes quantitatively precise relations between small eigenvalues of the generator associated to the dynamics and mean exit times of metastable domains. In the stochastic Ising model at low temperature, it yields sharp results that go far beyond the logarithmic equivalence obtained by the pathwise approach [19]. This technique was also used to sharpen the previously obtained results on the Kawasaki dynamics in the low temperature regime in a finite box [13] and to prove the first rigorous results in the case of growing boxes [14].

Beside spin systems at very low temperature, the investigation in the dynamical behavior of *disordered mean field systems* is of particular interests. One of the simplest model, from a static point of view, that has been studied intensively in the past, is the random field Curie–Weiss model at finite temperatures. Mathieu and Picco [85] and Fontes, Mathieu and Picco [50] first analyzed the long-time behavior of this model in the case where the random field can take only two values $\pm\varepsilon$. By using spectral methods, they obtained the leading exponential asymptotics for metastable exit times. Based on the potential theoretic approach Bovier, Eckhoff, Gayrard and Klein [15] improved the previous results by establishing sharp estimates of transition times between metastable states. They consider the case where the distribution of the random field is discrete but assumes finitely many values. As long as the random field assumes arbitrary values in a finite set, the dynamical behavior of this model can be described effectively by a Markovian dynamics on a lower-dimensional state space. Finally, in two recent papers [6, 5], Bianchi, Bovier and Ioffe analyzed first the metastable behaviour of the Curie-Weiss model at finite temperature in the case of random fields with continuously distributions. By exploiting a dual variational representation of capacities in terms of unit flows, due to Berman and Konsowa [4], they derived in [6] first sharp estimates on averaged metastable exit times including an explicit expression for the prefactor. Secondly, by means of coupling techniques, they showed in [5] that mean metastable exit times are almost constant as functions of the starting configuration within well chosen sets. Additionally, they proved the convergence of the law of normalized metastable exit times to an exponential distribution.

An outline of this book

By now, we have reached a fairly comprehensive understanding of metastable phenomena in at least two different settings: stochastic dynamics of spin systems in the low temperature regime and mean field models at finite temperature. A crucial feature of models at very low temperature is that i) metastable states correspond to local minima of the Hamiltonian and ii) the entropy of paths does not play a role, i.e. the transition between metastable states A, B is realized by microscopic paths in a arbitrary small tube around the optimal microscopic path that is given as the minimizer in the variational problem $\inf_{\gamma: A \to B} I(\gamma)$. On the other hand, in stochastic spin systems at finite temperature the entropy of paths really matters. Namely, not only individual microscopic paths have a vanishing small probability but their probability of stay, say, in a tube with radius $\delta > 0$ around a given microscopic path is very small. For this reason, one has to lump enormously many paths together in order to be able to describe a transition between metastable states. This is usually done by introducing a macroscopic variable and analyzing the dynamical properties of the resulting process. A characteristic feature of mean field models is that the induced dynamics on the *coarse-grained level* is still Markovian. In such a case, i) metastable states correspond to local minima of the free energy. More important is the fact that ii) by means of the macroscopic variable the dimension of the state space underlying the induced dynamics is diminished, i.e. it remains to study essentially the metastable behavior of a random walk in the free energy landscape. To some extend, this reduction procedure leads to a setting comparable with a low temperature regime where the role of the temperature, however, is replaced by the volume.

The purpose of this book is to study metastability for a class of stochastic spin systems at *finite temperature* which are *not* exactly reducible to a low-dimensional model via lumping techniques. Our main objective is to *extend* the potential theoretic approach to metastability for such kind of models by identifying key ingredients for deriving sharp estimates on

- the expected time of a transition from a metastable to a stable state,
- the distribution of the exit time from a metastable state,
- small eigenvalues of the generator.

While the first four chapters of this thesis focus on the general concepts of this approach, a real test of these ideas comes with chapter 5, where we apply this approach to a simple disordered spin system.

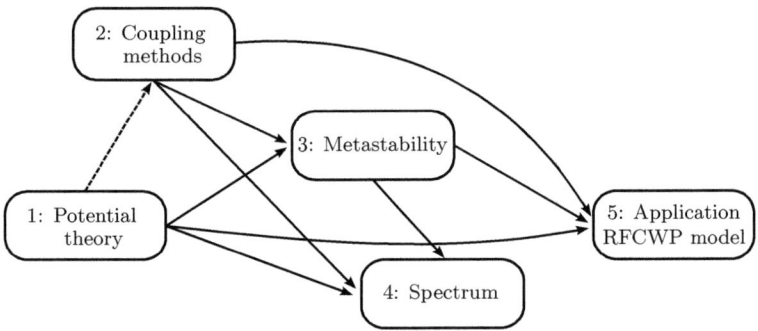

Figure 1. Logical dependence among the chapters

Chapter 1 is concerned with some aspects of the profound connection between Markov processes and potential theory that serves as a basis for the potential approach to metastability. The main idea of this approach is to express probabilistic objects, e.g. mean hitting times, first in terms of solutions of boundary value problems. Instead of trying to solve approximately the resulting equations, the second idea is to characterize the solution in terms of capacities and equilibrium potentials. Notice that in the setting, we are interested in, the formula for the mean hitting time of a set B when starting from A is only useful if A, B are sufficiently large. However, in such a situation, strict renewal equation breaks down that are crucial to control the equilibrium potential. One of the main objectives in the first chapter is the derive various averaged renewal equations that will be important in later chapters. In order to emphasis the link between probability and potential theory, we give for every statement both a probabilistic and two different analytic proofs.

Chapter 2 is mainly dedicated to establish pointwise estimates of various probabilistic objects. In situations where the process returns very often to the starting set A before escaping to the set B, one expects that the mean hitting time of B is almost constant as a function of a starting configuration $\sigma \in A$. In a recent paper [5], Bianchi, Bovier and Ioffe addressed this question for a class of Markov chains with state space $\{-1,+1\}^N$ by means of coupling and splitting techniques. The main objective in the second chapter is to generalize the coupling construction to a class of Markov chains with state space $\{1,\ldots,q\}^N$. In particular, we demonstrate that the splitting techniques, introduced in [5], are rather universal applicable. As a byproduct, we obtain rough estimates on the oscillations of harmonic functions within certain sets. Let us point out that these estimates are the second key ingredient in order to derive sharp bounds on mean hitting times.

In Chapter 3 we introduce the notion of a set of metastable sets. This definition is a natural generalization of the set of metastable points, first introduced in [15], and will be the starting point for our further investigations. Notice that in the context of reversible diffusions a similar generalization of metastable points was already discussed in [10]. One advantage of the potential theoretic approach to metastability is that it suffices to establish sharp bounds on capacities and to derive rough bounds on the corresponding equilibrium potentials in order to compute precisely metastable exit times, see [9, 11]. The later is previously done by means of exact renewal equations. Equipped with the results obtained in the first two chapters we demonstrate how averaged renewal equations combined with the rough estimates on the regularity of harmonic functions can be used to compute precisely the mean exit time from a metastable set and to prove the convergence of normalized metastable exit times to an exponential distribution. Notice that similar ideas were already used in [17] and [5].

In Chapter 4 we investigate the relation between small eigenvalues of the generator and mean hitting times of certain metastable sets. Previously, in the study of reversible diffusions on $\mathbb{R}^d$ or Markov chains on finite state spaces, a sharp link between metastable points and small eigenvalues of the generator could be established [18, 16]. Let us stress the fact that the method used in [18] is based on regularity properties of eigenfunctions. However, in the setting we are interested in, such sufficient control of the oscillations of eigenfunctions within certain sets is still be missing. For this reason, we invent a different approach that is based on an a posteriori error analysis and pointwise estimates of equilibrium potentials. Assuming that there are at most as many low-lying eigenvalues as metastable sets and taking a non-degeneracy condition into account, this approach allows to associate to each metastable set one simple eigenvalue of the generator. Moreover, every such eigenvalue is equal to the inverse of the mean exit time from this set up to exponentially small errors.

As an application of the theory developed above, we study in the final Chapter 5 the dynamical behavior of the *random field Curie-Weiss-Potts model* at finite temperature and with a continuous distribution of the random field. This model belongs to the class of disordered mean-field spin systems that cannot be reduced exactly to a lower-dimensional system. Since the free energy landscape in the thermodynamic limit is deterministic, from a static point of view, is model is one of the less disordered. The author's contribution in this chapter is to generalize the results that were previously obtained in the study of the random-field Curie-Weiss model, see [6].

Acknowledgements

This book is based on my PhD thesis. I warmly thank all who supported the writting of the thesis, in particular my supervisor Anton Bovier and my parents.

Throughout the research on my thesis, I have received financial support in part from the Weierstrass Institute for Applied Analysis and Stochastics (WIAS) in Berlin, the Institute of Applied Mathematics (IAM) at Bonn University, the German Research Council through the SFB 611 "Singuläre Phänomene und Skalierung in mathematischen Modellen", the International Research Trainings Group "Stochastic Models of Complex Processes" and the Bonn International Graduate School (BIGS). The kind hospitality of both the WIAS and IAM is gratefully acknowledged.

Berlin, July 2012 *Martin Slowik*

Chapter 1

Basic ingredients of the potential theoretic approach

In this first chapter, we present some aspects of the profound connection between Markov processes and potential theory. These serve as a basis for the *potential theoretic approach to metastability* developed systematically by Bovier, Eckhoff, Gayrard and Klein in [15, 16]. In later applications, we are mainly interested in the metastable behavior of processes coming from statistical mechanics. Typically, the state space for such kind of processes is large but finite. For this reason, we focus our attention to the study of stochastic processes on *finite state spaces* and in *discrete time* throughout this thesis. Let us emphasize that the potential theoretic approach to metastability is neither limited to finite state spaces nor to processes in discrete time. For instance, in the study of local Kawasaki dynamics in finite volumes with open boundary conditions [13] and in volumes that are growing to infinity [14] this approach was successfully applied to Markov chains in continuous time. In the context of diffusion processes in $\mathbb{R}^d$, the first rigorous proof of the classical *Eyring-Kramers formula* [17] and the precise control of the small eigenvalues of the associated generator [18, 45] rely as well on this method.

This chapter is organized as follows. In Section 1.1, after giving a brief introduction to classical potential theory and explaining its relation to probability theory, we specify the setting that is actually used in later applications. In Section 1.2, we treat the Dirichlet problem of finding a function which solves a Feynman-Kac-Poisson equation in a given region and assumes the values of a prescribed function at the boundary.

Markov chains can be used to characterize the solution of such a Dirichlet problem and to prove existence and uniqueness. Various applications of this probabilistic representation are given in Section 1.3. A different way to represent the solution of a Dirichlet problem is in terms of the Green function that will be introduced in Section 1.4. As presented in Section 1.5, one major achievement of the potential theoretic approach is the observation that instead of searching for an approximate solution to a Dirichlet problem, the solution can alternatively be represented in terms of capacities which satisfy variational principles. As we will see in Section 1.6 and 1.7, the formulation in terms of dual variational principles leads to the derivation of upper and lower bounds for the solution of the Dirichlet problem. The final Sections 1.8 and 1.9 deal with averaged versions of renewal equations and the almost ultrametricity of capacities which will be important in later chapters.

1.1. Introduction

One of the fundamental problems in potential theory is the classical Dirichlet problem. It originates from the study of physical questions coming from electrostatics. The investigation in existence and uniqueness of solutions to it has a long and venerable history which can be traced back e.g. to the work of Dirichlet [**38**], Poincaré [**98**], Kellogg [**75**], Lebesgue [**80**]. Various analytic methods have been invented to construct a solution: Schwarz alternating method, variational methods to name only two of them. The idea of Schwarz's alternating method is to decompose first the domain into two or more sub domains having a non-empty intersection. In a second step, similar Dirichlet problems are solved alternately on these sub domains, those values on the additional boundaries are given by the previous solution. In comparison to the Schwarz alternating method, the variational method is based on the fact that the solution of the Dirichlet problem assuming certain values at the boundary is also given as the minimizer of an appropriate energy functional.

Notice that, in general, the Dirichlet problem is not always solvable without assuming a certain regularity of the domain's boundary. This fact was observed at an early stage by considering the example of an punctured sphere or a sharp inward-pointing cups (Lebesgue's thorn).

A first probabilistic solution to the classical Dirichlet problem was given by Kakutani [**72**]. His pioneering work and the extensive study by Doob [**41**] and Hunt [**67, 68, 69**] established the profound connection between potential theory and Markov processes. To some extend these two theories are mathematically equivalent. This allows to translate a proof of a result in one theory into a proof of a corresponding result in the other theory, which we will demonstrate in Section 1.5 and 1.8. There is now a large literature on (transient) potential theory and Markov processes, including books e.g. by Blumenthal and Getoor [**8**], Port and Stone [**101**], Karatzas and Shreve [**73**, chap. 4] and the comprehensive treatment by Doob [43]. A nice introduction into this topic can be found in [102] and [27].

In the context of Markov chains on finite or denumerable state spaces, the connection to (recurrent) potential theory was developed by Doob [42] and Kemeny, Knapp and Snell [76]. While classical potential theory had already been well understood before its connection to Brownian motion was discovered, Markov chains and discrete potential theory were more or less simultaneously studied. Historically, in his famous paper "Random walks and electric currents in networks" [88], Nash-Williams first linked the property of transience or reccurence of an irreducible reversible Markov chain to structural properties of the underlying electric network. While Pólya [100] first proved by using purely probabilistic arguments that a simple random walk on $\mathbb{Z}^d$ returns to its starting position with probability 1 if and only if $d \leq 2$, Nash-Williams showed that this reccurence holds on graphs if and only if the effective resistance between the starting position and infinity of the corresponding electric network is infinite. Although the discrete potential theory has proven to be a robust tool to study properties of Markov chains, its application attracted new attention with the beautiful elementary book by Doyle and Snell [44]. Since that time various aspects of Markov chains have been studied exploiting the interdependence between probabilistic objects and its analytic counterparts and there is now a large literature on this topic including books e.g. by Woess [114], Telcs [108], Levin, Peres and Wilmer [82], Lyons and Peres [83].

The potential theoretic approach to metastability is based on the observation that most quantities of physical interest can be represented as solutions of certain *Dirichlet problems* with respect to the generator of the dynamics. More important is the fact that the corresponding solutions can be expressed in terms of *capacities* which satisfy *variational principles*. With proper physical insight, these variational principles allow us to derive reasonable upper and lower bounds on capacities.

Since various notions used in the potential theoretic approach are borrowed from electrostatics we will now explain briefly their physical background. For more details, we refer to the classical physical literature on this subject, see [71, 106].

Electrostatic interpretation. Many problems in electrostatics appearing in practical applications concern finite regions of space, i.e. bounded connected open subsets $D \subset \mathbb{R}^d$ having a smooth boundary ∂D with surface measure S, with or without charges inside D and with predescribed values given by a function on the bounding surfaces ∂D. The connection between a distribution of free charges $q(\mathrm{d}x)$ inside D and the values of the electric displacement $\boldsymbol{D}$ on ∂D is given by Gauss's law. The latter states that the flux of the electric displacement across the boundary is proportional to the sum of free charges enclosed by D. Hence, by Gauss' divergence theorem,

$$\int_D \operatorname{div} \boldsymbol{D}(x)\, \mathrm{d}x = \int_{\partial D} \langle \boldsymbol{D}(x), \boldsymbol{n}(x) \rangle\, S(\mathrm{d}x) = \int_D q(x)\, \mathrm{d}x, \qquad (1.1)$$

where we assume that the signed measure $q(\mathrm{d}x)$ has a density, i.e. $q(\mathrm{d}x) = q(x)\,\mathrm{d}x$. Here, $\boldsymbol{n}(x)$ denotes the outwardly directed surface normal of unit length for the surface ∂D at x. In the absence of a dielectric material, the constitutive relation between the electric field $\boldsymbol{E}$ and the electric displacement $\boldsymbol{D}$ is given by $\boldsymbol{D}(x) = \varepsilon_0\, \boldsymbol{E}(x)$, where $\varepsilon_0 > 0$ is the permittivity of free space.

On the other hand, Faraday's law of induction in its differential form relates the curl of the electric field to the time rate of change of the magnetic flux that vanishes in a static situation. Thus, this leads to the Maxwell equations

$$\operatorname{div} \boldsymbol{D}(x) = q(x) \quad \text{and} \quad \operatorname{rot} \boldsymbol{E}(x) = 0, \qquad x \in D \qquad (1.2)$$

which imply that there exists a potential $\varphi : \overline{D} \to \mathbb{R}$ such that $\boldsymbol{E} = -\nabla \varphi$. One of the common problems in electrostatic arising from (1.2) is the determination of the potential φ in a vacuum given a charge distribution q inside of D and a function u specifying the values of the potential at the boundary, i.e. to solve the Dirichlet problem

$$\begin{cases} \Delta \varphi(x) = -\frac{1}{\varepsilon_0}\, q(x), & x \in D \\ \varphi(x) = u(x), & x \in \partial D. \end{cases} \qquad (1.3)$$

Originally derived by Green [59] in 1828, an immediate consequence of the divergence theorem applied to $\varphi \nabla \psi$, where $\varphi, \psi \in C^2(\overline{D})$, is *Green's first identity*

$$\int_D \left(\varphi(x) \Delta \psi(x) + \langle \nabla \varphi(x), \nabla \psi(x) \rangle \right) \mathrm{d}x = \int_{\partial D} \varphi(x)\, \partial_n \psi(x)\, S(\mathrm{d}x),$$

where $\partial_n \varphi(x) = \langle \nabla \varphi(x), \boldsymbol{n}(x) \rangle$ denotes the normal derivative of φ on ∂D. *Green's second identity* is obtained by interchanging φ and ψ and subtracting, i.e.

$$\int_D \left(\varphi(x) \Delta \psi(x) - \psi(x) \Delta \varphi(x) \right) \mathrm{d}x = \int_{\partial D} \left(\varphi(x)\, \partial_n \psi(x) - \psi(x)\, \partial_n \varphi(x) \right) S(\mathrm{d}x).$$

Resulting from Green's second identity, an integral representation of the solution to (1.3) is given by

$$\varphi(x) = \frac{1}{\varepsilon_0\, \omega_{d-1}} \int_D q(z)\, G_D(x,z)\, \mathrm{d}z - \frac{1}{\omega_{d-1}} \int_{\partial D} u(z)\, \partial_n G_D(x,z)\, S(\mathrm{d}z) \qquad (1.4)$$

where $\omega_{d-1} = 2\pi^{d/2}/\Gamma(\frac{d}{2})$ is the surface area of a $(d-1)$-sphere in $\mathbb{R}^d$. $G_D(x,y)$ is called the Green function that is, for any $y \in \overline{D}$, the fundamental solution in distributional sense of $\Delta G_D(x,y) = -\omega_{d-1}\, \delta(x-y)$ on D that vanishes on ∂D.

Of particular interest is the situation where the boundary of the region D consists of two separate conducting surfaces ∂A and ∂B which are hold at potential $\varphi_A = 1$ and

1.2. Boundary value problems

$\varphi_B = 0$, respectively, while the region D is free of electrical charges. The *capacity*, C, of this capacitor is defined by $C := Q_A/(\varphi_A - \varphi_B) = Q_A$ where

$$Q_A = -\varepsilon_0 \int_{\partial A} \langle \boldsymbol{E}(x), \boldsymbol{n}(x) \rangle\, S(\mathrm{d}x) = \varepsilon_0 \int_{\partial A} \partial_n \varphi(x)\, S(\mathrm{d}x) \qquad (1.5)$$

is the total charge on ∂A. Notice that Green's first identity immediately gives rise to the following representation of the capacity as a quadratic form

$$\begin{aligned}
C = Q_A &= \varepsilon_0\, \varphi_A \int_{\partial A} \partial_n \varphi(x)\, S(\mathrm{d}x) + \varepsilon_0\, \varphi_B \int_{\partial B} \partial_n \varphi(x)\, S(\mathrm{d}x) \\
&= \varepsilon_0 \int_{\partial D} \varphi(x)\, \partial_n \varphi(x)\, S(\mathrm{d}x) - \varepsilon_0 \int_D \varphi(x) \Delta \varphi(x)\, \mathrm{d}x \\
&= \varepsilon_0 \int_D \|\nabla \varphi(x)\|^2\, \mathrm{d}x.
\end{aligned}$$

The advantage of this representation is that it allows to establish dual variational principles as we will see in Section 1.6.

Setting. In what follows, let $(\Omega, \mathcal{F}, \mathbb{P})$ be a probability space and $\{\mathcal{F}_t\}_{t\in\mathbb{N}_0}$ be an increasing sequence of sub-σ-algebras of $\mathcal{F}$, which are fixed. Further, let $(\mathcal{S}_N, \mathcal{B}_N)$ be a measurable space where $\mathcal{S}_N$ is assumed to be a finite set. The cardinality of $\mathcal{S}_N$ depends on an addition parameter N and will diverge as N tends to infinity. Elements of $\mathcal{S}_N$ are denoted by Greek letters σ, η and called configurations. We consider a Markov chain $\{\sigma(t)\}_{t\in\mathbb{N}_0}$ on $\mathcal{S}_N$ with transitions probabilities $p_N(\sigma, \eta)$. Assume that the dynamics is *irreducible* and *reversible* with respect to a unique invariant measure μ_N. In particular, the transition probabilities satisfy the detailed balance condition $\mu_N(\sigma)\, p_N(\sigma, \eta) = \mu_N(\eta)\, p_N(\eta, \sigma)$ for all $\sigma, \eta \in \mathcal{S}_N$. By L_N we denote the generator which acts on functions $f \colon \mathcal{S}_N \to \mathbb{R}$ as

$$(L_N f)(\sigma) = \sum_{\eta \in \mathcal{S}_N} p_N(\sigma, \eta)\, \bigl(f(\eta) - f(\sigma)\bigr). \qquad (1.6)$$

Further, we denote by $\mathbb{P}_\nu$ the law of the Markov chain given that it starts with initial distribution ν and by $\mathbb{E}_\nu$ the expectation with respect to $\mathbb{P}_\nu$. If the initial distribution is concentrated on a single configuration η, we write simply $\mathbb{P}_\eta$ and $\mathbb{E}_\eta$, respectively. For any $A \subset \mathcal{S}_N$, let τ_A be the first hitting time of the set A *after* time zero, i.e.

$$\tau_A := \inf \bigl\{ t > 0 \bigm| \sigma(t) \in A \bigr\}.$$

When the set A is a singleton $\{\eta\}$ we write simply τ_η instead of $\tau_{\{\eta\}}$.

1.2. Boundary value problems

In this section, we turn our attention to the discrete analog of the classical *Dirichlet boundary value problem*. That is, given a non-empty subset $D \subset \mathcal{S}_N$, measurable functions $g, k \colon D^c \to \mathbb{R}$ and a measurable function $u \colon D \to \mathbb{R}$, the task is to find a (bounded) function $f \colon \mathcal{S}_N \to \mathbb{R}$ which satisfies

$$\begin{cases} (L_N f)(\sigma) - k(\sigma) f(\sigma) = -g(\sigma), & \sigma \in D^c \\ f(\sigma) = u(\sigma), & \sigma \in D, \end{cases} \quad (1.7)$$

where $D^c = \mathcal{S}_N \setminus D$. Provided that such a function f exists, it will be called *solution to the Dirichlet problem*.

While the probabilistic solution of (1.7) in the discrete-time setting is well established in cases where the function k vanishes or is identical to a constant, see for instance [91], we could not find a reference dealing with the probabilistic representation in the discrete-time context for general function k, although we expect that it exists. For this reason and for later reference, we will give below a proof of the corresponding statement.

Based on the deep connection between Markov processes and martingales, the probabilistic method allows to write down immediately a very likely candidate for a solution of (1.7). While uniqueness is obtained by applying Doob's optional stopping theorem to a suitable martingale, the proof of existence is in general more subtle, since the smoothness of a solution nearby and at the boundary of the domain depends itself on the regularity of the boundary. In the context of a discrete-time Markov chain on a finite state space the problem concerning the regularity of a solution at the boundary of the domain is not an issue and the existence can be easily achieved by using the Markov property.

To start with, consider a Markov chain $\{\sigma(t)\}$ in discrete-time with generator L_N. Then, for all (bounded) measurable functions f, it holds that

$$f(\sigma(t)) - \sum_{s=0}^{t-1} (L_N f)(\sigma(s)) \quad (1.8)$$

is a martingale with respect to $\mathbb{P}_\sigma$ for all $\sigma \in \mathcal{S}_N$. The following lemma provides us a useful equivalence to (1.8) being a martingale.

Lemma 1.1. *Let $\{\mathcal{F}_t\}$ be a filtration such that the Markov chain $\{\sigma(t)\}$ on $\mathcal{S}_N$ is adapted. Let $f, k \colon \mathcal{S}_N \to \mathbb{R}$ be (bounded) measurable functions. Further suppose that the function k satisfies $\min_{\eta \in \mathcal{S}_N} k(\eta) > -1$. Then, for all $\sigma \in \mathcal{S}_N$, (1.8) is a $\mathbb{P}_\sigma$-martingale if and only if*

$$M_t = f(\sigma(t)) \prod_{s=0}^{t-1} \frac{1}{1 + k(\sigma(s))}$$
$$+ \sum_{s=0}^{t-1} \Big(k(\sigma(r)) f(\sigma(r)) - (L_N f)(\sigma(r)) \Big) \prod_{r=0}^{s} \frac{1}{1 + k(\sigma(r))} \quad (1.9)$$

is a $\mathbb{P}_\sigma$-martingale.

1.2. Boundary value problems

Here we use the convention that empty products, i.e. products where the upper index is smaller than the lower one, are equal 1 and empty sums are equal 0.

Proof. In order to simplify notations and for later reference, we set

$$X_t := f(\sigma(t)) - \sum_{s=0}^{t-1} (L_N f)(\sigma(s)) \quad \text{and} \quad V_t := \prod_{s=0}^{t-1} \frac{1}{1+k(\sigma(s))}$$

Notice that the process $\{V_t\}$ is previsible and, due to the assumption on k, locally bounded. Thus, for all $\sigma \in S_N$, the theorem of discrete stochastic integrals implies that $\{X_t\}$ is a $\mathbb{P}_\sigma$-martingale if and only $\{(V \bullet X)_t\}$ is a $\mathbb{P}_\sigma$-martingale. On the other hand, a straight forward computation shows that

$$\begin{aligned}
(V \bullet X)_t &= \sum_{s=1}^{t} V_s (X_s - X_{s-1}) \\
&= \sum_{s=1}^{t} V_s \Big(f(\sigma(s)) - f(\sigma(s-1)) - (L_N f)(\sigma(s-1)) \Big) \\
&= V_t f(\sigma(t)) - V_1 f(\sigma(0)) \\
&\quad + \sum_{s=1}^{t-1} (V_s - V_{s+1}) f(\sigma(s)) - \sum_{s=0}^{t-1} V_{s+1} (L_N f)(\sigma(s)) \\
&= M_t - f(\sigma(0)). \quad (1.10)
\end{aligned}$$

Since $f(\sigma(0))$ is a constant under $\mathbb{P}_\sigma$, we conclude the proof of this lemma. $\square$

The martingale (1.9) is the key ingredient to prove existence and uniqueness.

Proposition 1.2. Let $k : D^c \to [\kappa, \infty)$ be a bounded function with $\kappa > -1$. Further assume that for some $\sigma \in D^c$

$$\frac{1}{1+\kappa} \cdot \limsup_{s \to \infty} (p_D^s(\sigma, D))^{1/s} < 1 \quad (1.11)$$

where $p_D(\sigma, \eta)$ denotes the transition probabilities of a sub-Markov chain that is stopped at the arrival of D. Then,

$$f(\sigma) = \mathbb{E}_\sigma \left[u(\sigma(\tau_D)) \prod_{s=0}^{\tau_D - 1} \frac{1}{1+k(\sigma(s))} + \sum_{s=0}^{\tau_D - 1} g(\sigma(s)) \prod_{r=0}^{s} \frac{1}{1+k(\sigma(r))} \right] \quad (1.12)$$

for every $\sigma \in D^c$ and $f \equiv u$ on D is a solution of the Dirichlet problem (1.7). In particular, the solution is unique.

Remark 1.3. First of all notice that if (1.11) is satisfied for some $\sigma \in D^c$ and $\eta \in D$, then, by irreducibility, it holds simultaneously for all $\sigma \in D^c$ and $\eta \in D$. Moreover, provided that the function k has the property that $k(\eta) > 0$ for all $\eta \in D^c$ then (1.11) is immediately satisfied.

Proof. First, let us consider a function f that solves (1.7), i.e. $(L_N f) - kf = -g$ on D^c. Further, it is seen from Lemma 1.1 that

$$M_t^{\tau_D} := f(\sigma(t \wedge \tau_D)) \prod_{s=0}^{t \wedge \tau_D - 1} \frac{1}{1 + k(\sigma(s))} + \sum_{s=0}^{t \wedge \tau_D - 1} g(\sigma(s)) \prod_{r=0}^{s} \frac{1}{1 + k(\sigma(r))}$$

is a local martingale up to time τ_D. Given a localizing sequence $\tau_D^n \uparrow \tau_D$ of stopping time, we construct another localizing sequence by

$$\tau_n := \tau_D^n \wedge \inf\{t > 0 \mid V_t > n\}. \tag{1.13}$$

This implies that for each n there exists $K_n < \infty$ such that

$$V_{\tau_n} = \frac{V_{\tau_n - 1}}{1 + k(\sigma(\tau_n - 1))} \leq \max_{\eta \in D^c} \frac{n}{1 + k(\eta)} \leq K_n, \tag{1.14}$$

where we exploit the assumption on the function k. In view of the martingale transform (1.10), (1.14) allows us to bound

$$\left| M_t^{\tau_n(\omega)} - M_{t-1}^{\tau_n(\omega)} \right| \leq 2 \max_{\eta \in S_N} |f(\eta)| K_n < \infty$$

uniformly for all $t \in \mathbb{N}_0$ and $\omega \in \Omega$. Since the dynamics is assumed to be irreducible and the state space is finite, it holds that $\mathbb{E}_\sigma[\tau_D] < \infty$. Hence, an application of the optional stopping theorem to the localization sequence τ_n reveals

$$f(\sigma) = \mathbb{E}_\sigma \left[u(\sigma(\tau_n)) \prod_{s=0}^{\tau_n - 1} \frac{1}{1 + k(\sigma(s))} + \sum_{s=0}^{\tau_n - 1} g(\sigma(s)) \prod_{r=0}^{s} \frac{1}{1 + k(\sigma(r))} \right], \tag{1.15}$$

for all $\sigma \in D^c$. The assertion (1.12) now follows provided we can interchange the limit $n \to \infty$ and the expectation value. Notice that the random variable involved in the expectation (1.15) is bounded from above by

$$\max_{\eta \in D} |u(\eta)| + \tau_D(\omega) \cdot \max_{\sigma \in D^c} |g(\eta)|$$

or

$$\left(\max_{\eta \in D} |u(\eta)| + \frac{1}{\kappa} \cdot \max_{\sigma \in D^c} |g(\eta)| \right) (1 + \kappa)^{-\tau_D(\omega)},$$

for $\mathbb{P}_\sigma$-a.a. $\omega \in \Omega$ depending on whether $\kappa \geq 0$ or $\kappa < 0$. In the later, due to the irreducibility and the assumption (1.11)

$$\mathbb{E}_\sigma\left[(1+\kappa)^{-\tau_D}\right] = \sum_{s=1}^{\infty} (1+\kappa)^{-s} p_D^s(\sigma, D) < \infty, \qquad \forall \sigma \in D^c. \tag{1.16}$$

Hence, we can apply Lebesque's dominated convergence theorem. This proves the uniqueness of representation (1.12). On the other hand, by using the Markov property,

1.2. Boundary value problems

a straight forward computation shows that for any $\sigma \in D^c$

$$(P_N f)(\sigma) = \mathbb{E}_\sigma\left[\mathbf{1}_{\tau_D \leq 1} u(\sigma(\tau_D))\right]$$

$$+ (1 + k(\sigma)) \mathbb{E}_\sigma\left[\mathbf{1}_{\tau_D > 1} \mathbb{E}_{\sigma(1)}\left[u(\sigma(\tau_D)) V_{\tau_D} + \sum_{s=1}^{\tau_D - 1} g(\sigma(s)) V_{s+1}\right]\right]$$

$$= (1 + k(\sigma)) f(\sigma) - g(\sigma).$$

This complete the proof. □

Remark 1.4. Concerning the function k, a natural question to ask is what is the reason behind the fact that the probabilistic representation (1.12) is infinite as soon as $k(\sigma) = -1$ for some $\sigma \in D^c$. A closer look at (1.11) and (1.16) reveals that the finiteness of (1.12) is related to the fact that κ is within the radius of convergence of the corresponding Neumann series. In order to illustrated this fact, consider for simplicity the Dirichlet problem for vanishing u and constant function k. We denote by L_D the restriction of the generator to D^c, i.e. $L_D(\sigma, \eta) = L_N(\sigma, \eta)$ for $\sigma, \eta \in D^c$ and by P_D the corresponding restriction of the transition matrix P_N. If $L_D - k \cdot I$ is regular, the analytic solution to (1.7) is given by $f = (k \cdot I - L_D)^{-1} g$. On the other hand, provided that $\frac{1}{1+k}\|P_D\| < 1$, we can represent the inverse in terms of a Neumann series, i.e.

$$(k \cdot I - L_D)^{-1} = \tfrac{1}{1+k}\left(I - \tfrac{1}{1+k} P_D\right)^{-1} = \sum_{s=0}^{\infty} \left(\tfrac{1}{1+k}\right)^{-(s+1)} P_D^s \qquad (1.17)$$

The right-hand side of (1.17) correspond to the probabilistic representation, see (1.31).

Remark 1.5. Let us briefly comment on the similarities and differences in the probabilistic representation of a solution to the Dirichlet problem in discrete-time compared to the one in continuous-time. The latter in given by

$$f(\sigma) = \mathbb{E}_\sigma\left[u(\sigma(\tau_D)) e^{-\int_0^{\tau_D} k(\sigma(s))\, ds} + \int_0^{\tau_D} g(\sigma(s)) e^{-\int_0^s k(\sigma(r))\, dr}\, ds\right]. \qquad (1.18)$$

If $k \geq 0$ and $L_N \equiv \tfrac{1}{2}\Delta$, the Feyman-Kac formula (1.18) can be interpreted as a Brownian motion with killing of particles at the same rate k that is stopped at the arrival of the boundary of D. In particular, the probability that a particle survives up to time t, conditioned on the path $\{\sigma(s)\}_{0 \leq s \leq t}$, is equal to $\exp(-\int_0^t k(\sigma(s))\, ds)$. In discrete-time, (1.12) suggests that the situation is equivalent to a *random walk with killing* where the conditional survival probability of a particle up to time t is then $\prod_{s=0}^{t} 1/(1 + k(\sigma(s)))$.

In the case when $-1 < k \leq 0$, (1.12) can be seen as a random walk of particles that *branches of* with the same rate k, and an analog interpretation can be given to the Feyman-Kac formula in the continuous-time setting. For obvious reasons, the solution may explosion in finite time such that we have to impose conditions on the function k. Further we want to stress that if $k < -2$, the probabilistic representation (1.12) is finite and has no counterpart in the continuous-time setting.

Remark 1.6. Consider the case that the function k vanishes on D^c. Then the Dirichlet problem (1.7) is known as *Laplace equation* and *Poisson equation*, respectively, depending on whether either $g \equiv 0$ or $u \equiv 0$. In the case $k \not\equiv 0$, (1.7) is also called *Feynman-Kac* or *Feynman-Kac-Poisson equation*.

1.3. Laplace equation and Poisson equation

Proposition 1.2 is a crucial element in the theory since it allows us to compute many interesting expectation values for Markov chains by solving an appropriate boundary value problem. In the sequel, we will consider various corresponding applications.

The key quantity in the potential theoretic approach is the λ-*equilibrium potential*, $h_{A,B}^\lambda$. In order to define it, consider a *capacitor*, (A, B), build up by two disjoint subsets $A, B \subset \mathcal{S}_N$. Then, $h_{A,B}^\lambda$ is defined as the solution of the Dirichlet problem (1.7) with $g \equiv 0$ and $k \equiv e^\lambda - 1$, for $\lambda \geq 0$, subject to the boundary condition given by the indicator function $\mathbb{1}_A$ on the set A. The corresponding boundary value reads

$$\begin{cases} (L_N f)(\sigma) - \left(e^\lambda - 1\right) f(\sigma) = 0, & \sigma \in (A \cup B)^c \\ f(\sigma) = \mathbb{1}_A(\sigma), & \sigma \in A \cup B. \end{cases} \quad (1.19)$$

A related quantity is the λ-*equilibrium measure*, $e_{A,B}^\lambda$, which is defined through

$$e_{A,B}^\lambda(\sigma) := -(L_N^\lambda h_{A,B}^\lambda)(\sigma), \qquad \forall \sigma \in \mathcal{S}_N, \quad (1.20)$$

where we introduce the operator $L_N^\lambda := L_N - (e^\lambda - 1)$ to shorten notation. In case $\lambda = 0$, we simply write $h_{A,B}$ and $e_{A,B}$ instead of $h_{A,B}^0$ and $e_{A,B}^0$.

Clearly, the equilibrium measure is non-vanishing only on the boundary of the sets A and B. An immediate consequence of Proposition 1.2 is that the equation (1.19) has a unique solution for all $\lambda \geq 0$. Moreover, in view of (1.12), the function $h_{A,B}^\lambda$ describes the following probabilistic objects. Namely, in the case $\lambda > 0$ the λ-equilibrium potential, $h_{A,B}^\lambda$, represents the Laplace transform of the hitting time, τ_A, of the set A taking into account that the Markov chain $\{\sigma(t)\}$ starting in σ gets kill at the arrival of B. For $\lambda = 0$, the equilibrium potential has a natural interpretation in terms of hitting probabilities. Beside $h_{A,B}^\lambda$ the λ-equilibrium measure has as well a probabilistic representation that can be derived by using the Markov property. All in one, we have for $\lambda \geq 0$

$$\mathbb{E}_\sigma \left[e^{-\lambda \tau_A} \mathbb{1}_{\tau_A < \tau_B} \right] = \begin{cases} h_{A,B}^\lambda(\sigma), & \sigma \notin A \cup B, \\ 1 - e^{-\lambda} e_{A,B}^\lambda(\sigma), & \sigma \in A, \\ -e^{-\lambda} e_{A,B}^\lambda(\sigma), & \sigma \in B. \end{cases} \quad (1.21)$$

1.3. Laplace equation and Poisson equation

Notice that in the case $\lambda = 0$, by distinguish two cases, (1.21) implies that for all $\sigma \in A \cup B$

$$e_{A,B}(\sigma) = -(L_N h_{A,B})(\sigma) = (L_N h_{B,A})(\sigma) = -e_{B,A}(\sigma). \qquad (1.22)$$

A further object of interest is the Laplace transform of the hitting time of a set $D \subset S_N$, which is the solution of a Feynman-Kac equation similar to (1.19). More precise, if we denote by $w_D^\lambda = \mathbb{E}_\sigma[e^{-\lambda \tau_D}]$, for $\sigma \in D^c$, then w_D^λ is the solution of the Dirichlet problem (1.7) with $g \equiv 0$, $k \equiv e^\lambda - 1$, for $\lambda \geq 0$, subject to the boundary condition 1 on the set D.

Let us now turn our attention to the Poisson equation. If the functions k and u vanish, the boundary value problem (1.7) reduces to

$$\begin{cases} (L_N f)(\sigma) = -g(\sigma), & \sigma \in D^c \\ f(\sigma) = 0, & \sigma \in D. \end{cases} \qquad (1.23)$$

In view of Proposition 1.2, the solution of the Poisson equation has the representation

$$f(\sigma) = \mathbb{E}_\sigma\left[\sum_{s=0}^{\tau_D - 1} g(\sigma(s))\right], \quad \forall \sigma \in D^c, \qquad (1.24)$$

which can be interpreted as average cost accumulated along a path of the Markov chain, $\{\sigma(t)\}$, before hitting the set D. Of particular interest is the mean hitting time of the set D which is obtained by choosing $g \equiv 1$ on D^c. Denoting the corresponding solution by w_B we have

$$w_D(\sigma) = \mathbb{E}_\sigma[\tau_D], \quad \forall \sigma \in D^c. \qquad (1.25)$$

Let us now consider two disjoint subsets $A, B \subset S_N$. We will be interested in the mean hitting time of A starting the Markov chain in a configuration $\sigma \notin A \cup B$ that gets killed at the arrival of B. To start with, let us denote by $w_{A,B}$ the solution of the Poisson equation (1.23) for the choice $D = A \cup B$ and $g = h_{A,B}$. In this case (1.24) reads

$$w_{A,B}(\sigma) = \mathbb{E}_\sigma\left[\sum_{s=0}^{\tau_{A \cup B} - 1} h_{A,B}(\sigma)\right], \quad \forall \sigma \notin A \cup B. \qquad (1.26)$$

The following lemma shows that $w_{A,B}$ is indeed the object we are interested in.

Lemma 1.7. *Let $A, B \subset S_N$ be two disjoint subsets of the finite state space S_N. Then,*

$$w_{A,B}(\sigma) = \mathbb{E}_\sigma[\tau_A \mathbb{1}_{\tau_A < \tau_B}], \quad \forall \sigma \notin A \cup B. \qquad (1.27)$$

Proof. Recall that $h_{A,B}$ is harmonic in $(A \cup B)^c$. As a consequence of the discrete martingale problem $X_t := h_{A,B}(\sigma(t)) - \sum_{s=0}^{t-1}(L_N h_{A,B})(\sigma(s))$ is a martingale and it holds that $X_t = h_{A,B}(\sigma(t))$ as long as $t < \tau_{A \cup B}$ provided the Markov chain is started in $\sigma \notin A \cup B$. Since the process $H_t = t$ is previsible and locally bounded, $M_t := (H \bullet X)_t$

is a martingale that vanishes in zero. In particular, M_t is a local martingale up to time $\tau_{A\cup B}$. By applying the optional stopping theorem with a localizing sequence $\tau_n \uparrow \tau_{A\cup B}$ of stopping times we obtain

$$0 = \mathbb{E}_\sigma\left[(H \bullet X)_0\right] = \mathbb{E}_\sigma\left[(H \bullet X)_{\tau_n}\right] = \mathbb{E}_\sigma\left[\tau_n\, h_{A,B}(\sigma(\tau_n)) - \sum_{s=0}^{\tau_n - 1} h_{A,B}(\sigma(s))\right]. \quad (1.28)$$

Since $\mathbb{E}_\sigma[\tau_B] < \infty$, Lebesque's dominated convergence theorem implies that, for all $\sigma \notin A \cup B$,

$$w_{A,B}(\sigma) = \mathbb{E}_\sigma\left[\tau_{A\cup B}\, h_{A,B}(\sigma(\tau_{A\cup B}))\right] = \mathbb{E}_\sigma\left[\tau_A\, \mathbb{1}_{\tau_A < \tau_B}\right] \quad (1.29)$$

where we used that the $h_{A,B}(\sigma) = \mathbb{1}_A(\sigma)$ for $\sigma \in A \cup B$. This completes the proof. $\square$

1.4. Green function and Green's identities

Let us now focus on a specific Feynman-Kac-Poisson equation that is given by the Dirichlet problem (1.7) with the choice of vanishing function u and $k \equiv e^\lambda - 1$, for $\lambda \geq 0$, i.e.

$$\begin{cases} (L_N f)(\sigma) - \left(e^\lambda - 1\right) f(\sigma) = -g(\sigma), & \sigma \in D^c \\ f(\sigma) = 0, & \sigma \in D. \end{cases} \quad (1.30)$$

The stochastic representation (1.12) for a solution of (1.30) can be written in the form

$$f(\sigma) = \mathbb{E}_\sigma\left[\sum_{s=0}^{\tau_D - 1} e^{-\lambda(s+1)} g(\sigma(s))\right] = \sum_{\eta \notin D} G_D^\lambda(\sigma, \eta)\, g(\eta) \quad \forall\, \sigma \in D^c, \quad (1.31)$$

where $G_D^\lambda : \mathcal{S}_N \times \mathcal{S}_N \to \mathbb{R}$ is called the *Green function*, which is defined through

$$G_D^\lambda(\sigma, \eta) := \sum_{s=0}^{\infty} e^{-\lambda(s+1)}\, \mathbb{P}_\sigma\left[\sigma(s) = \eta,\, s < \tau_D\right], \quad \forall\, \sigma, \eta \in D^c \quad (1.32)$$

and

$$G_D^\lambda(\sigma, \eta) := \mathbb{1}_{\sigma = \eta} \quad \forall\, \sigma, \eta \in D, \quad \text{and} \quad G_D^\lambda(\sigma, \eta) := 0, \quad \text{otherwise.} \quad (1.33)$$

In the case $\lambda = 0$, the Green function represents the expected number of visits in η starting the Markov chain, $\{\sigma(t)\}$, in σ before hitting the set D. Moreover, we denote by

$$p_D^s(\sigma, \eta) := \mathbb{P}_\sigma\left[\sigma(s) = \eta,\, s < \tau_D\right] < p_N^s(\sigma, \eta) \quad (1.34)$$

the transition function of a process that is absorbed in D. Notice that $p_D^s(\sigma, \cdot)$ are subprobabilities distributions. Since the dynamics is assumed to be reversible, we have

$$\mu_N(\sigma)\, G_D^\lambda(\sigma, \eta) = \mu_N(\eta)\, G_D^\lambda(\eta, \sigma). \quad (1.35)$$

1.4. Green function and Green's identities

Notice that the λ-equilibrium potential has an immediate representation in terms of the Green function. Namely, for all $\sigma \in B^c$ it holds that

$$h^\lambda_{A,B}(\sigma) = \sum_{\eta \in A} G^\lambda_B(\sigma, \eta) \, e^\lambda_{A,B}(\eta). \tag{1.36}$$

For later reference, the representation of w_D via the Green function is given by

$$w_D(\sigma) = \sum_{\eta \notin D} G_D(\sigma, \eta). \tag{1.37}$$

An important formula is the discrete analog of the first and second Green's identity. Let us emphasis the fact that reversibility of the dynamics implies that the (discrete) generator, L_N, is self-adjoint on $L^2(\mathcal{S}_N, \mu_N)$. As a consequence, we have

Proposition 1.8 (First Green's identity). *Let $f, g \in L^2(\mathcal{S}_N, \mu_N)$. Then,*

$$\frac{1}{2} \sum_{\sigma, \eta \in \mathcal{S}_N} \mu_N(\sigma) \, p_N(\sigma, \eta) \left(f(\sigma) - f(\eta) \right) \left(g(\sigma) - g(\eta) \right)$$
$$= - \sum_{\sigma \in \mathcal{S}_N} \mu_N(\sigma) \, f(\sigma) \, (L_N g)(\sigma). \tag{1.38}$$

Proof. By expanding the left-hand side of (1.38) and using the detailed balance condition the assertion is immediate. □

Corollary 1.9 (Second Green's identity). *Let $A \subset \mathcal{S}_N$ and $f, g \in L^2(\mathcal{S}_N, \mu_N)$. Then,*

$$\sum_{\sigma \in A} \mu_N(\sigma) \left(f(\sigma) \, (L_N g)(\sigma) - g(\sigma) \, (L_N f)(\sigma) \right)$$
$$= \sum_{\sigma \in A^c} \mu_N(\sigma) \left(g(\sigma) \, (L_N f)(\sigma) - f(\sigma) \, (L_N g)(\sigma) \right). \tag{1.39}$$

Remark 1.10. In the context of reversible diffusion processes in $\mathbb{R}^d$, studied in [17, 18], Green's identities were crucial to derive various estimates. Emphasizing this link, we will refer to (1.39) as second Green's identity, although it is a simple rewriting of $\langle f, L_N g \rangle_{\mu_N} = \langle L_N f, g \rangle_{\mu_N}$ in the discrete setting.

Remark 1.11. While the discrete analogue of Green's first identity is well known and its application can be found at various places in the literature, the strength of Green's second identity, i.e. the systematic exploit of the reversibility, seems to be overlooked in the past.

We turn now our attention to the Feynman-Kac equation which is given by the Dirichlet problem (1.7) with vanishing function g and $k \equiv e^\lambda - 1$, for $\lambda \geq 0$. It reads

$$\begin{cases} (L_N f)(\sigma) - (e^\lambda - 1) \, f(\sigma) = 0, & \sigma \in D^c \\ f(\sigma) = u(\sigma), & \sigma \in D. \end{cases} \tag{1.40}$$

Our aim is to derive a representation of the solution operator, H_D^λ, associated to the Feynman-Kac equation in terms of the Green function to which we will refer to as *Poisson kernel*. In order to do so, suppose that f solves (1.30). Further, for any fixed $\sigma \in D^c$ an application of the second Green's identity to the functions f and $G_D^\lambda(\cdot, \sigma)$ reveals immediately that

$$f(\sigma) = \sum_{\eta \in D} \frac{\mu_N(\eta)}{\mu_N(\sigma)} \left(L_N^\lambda G_D^\lambda\right)(\eta, \sigma) u(\eta), \qquad (1.41)$$

where we used that, for all $\eta \in D^c$, $(L_N^\lambda G_D^\lambda)(\eta, \sigma) = -\mathbb{1}_{\eta=\sigma}$ and $(L_N^\lambda f)(\eta) = 0$. Recall that $L_N^\lambda = L_N - (e^\lambda - 1)$. In view of the probabilistic representation (1.12), we have

$$H_D^\lambda(\sigma, \eta) = \frac{\mu_N(\eta)}{\mu_N(\sigma)} \left(L_N^\lambda G_D^\lambda\right)(\eta, \sigma) = \mathbb{E}_\sigma\left[e^{-\lambda \tau_D} \mathbb{1}_{\sigma(\tau_D) = \eta}\right], \qquad \forall \sigma \in D^c, \eta \in D, \tag{1.42}$$

otherwise we set $H_D^\lambda(\sigma, \eta) = 0$. As an application, we obtain for the λ-equilibrium potential the following representation

$$h_{A,B}^\lambda(\sigma) = \sum_{\eta \in A} \frac{\mu_N(\eta)}{\mu_N(\sigma)} \left(L_N^\lambda G_{A \cup B}^\lambda\right)(\eta, \sigma), \qquad \forall \sigma \notin A \cup B. \tag{1.43}$$

1.5. Equilibrium potential, capacity, and mean hitting times

In this section, we will introduce *the* main object of the potential theoretic approach. Here, we restrict our attention to the case $\lambda = 0$. The *capacity* of the capacitor (A, B) with equilibrium potential one on A and zero on B is defined through

$$\mathrm{cap}(A, B) := \sum_{\sigma \in A} \mu_N(\sigma) e_{A,B}(\sigma). \tag{1.44}$$

Let us denote by $\langle \cdot, \cdot \rangle_{\mu_N}$ the scalar product in $L^2(\mathcal{S}_N, \mu_N)$. The *energy* associated to the pair (P_N, μ_N) is defined for any function f on $\mathcal{S}_N$ by

$$\mathcal{E}(f) := \langle -L_N f, f \rangle_{\mu_N}. \tag{1.45}$$

An immediate consequence of the first Green's identity is that

$$\mathcal{E}(f) = \frac{1}{2} \sum_{\sigma, \eta \in \mathcal{S}_N} \mu_N(\sigma) p_N(\sigma, \eta) \left(f(\sigma) - f(\eta)\right)^2. \tag{1.46}$$

Due to the detailed balance condition, the factor $1/2$ ensure that each pair of configuration (σ, η) contributes to the energy only once. Inspecting the definition of the capacity reveals that $\mathrm{cap}(A, B) = \mathcal{E}(h_{A,B})$. In particular, we obtain that

$$\mathrm{cap}(A, B) = \frac{1}{2} \sum_{\sigma, \eta \in \mathcal{S}_N} \mu_N(\sigma) p_N(\sigma, \eta) \left(h_{A,B}(\sigma) - h_{A,B}(\eta)\right)^2. \tag{1.47}$$

1.5. Equilibrium potential, capacity, and mean hitting times

Moreover, the capacity has a probabilistic interpretation in terms of *escape probabilities*, namely $\mathbb{P}_{\mu_A}[\tau_B < \tau_A] = \text{cap}(A,B)/\mu_N[A]$, where $\mu_A(\sigma) = \mu_N[\sigma|A]$, for $\sigma \in A$, stand for the reversible measure conditioned on the set A.

So far, we have seen that the mean hitting time, $\mathbb{E}_\sigma[\tau_B]$, of a set B, has an analytic counterpart, w_B, that solves a certain Poisson equation. However, in most applications an exact solution of the corresponding boundary value problem can not be achieved. The first important ingredient of the potential theoretic approach to metastability is a formula for the average mean hitting time when starting the Markov chain in the so called *last exit biased distribution* on a set A that connects it to the capacity and the equilibrium potential. For any two disjoint subsets $A, B \subset \mathcal{S}_N$ this distribution is defined through

$$\nu_{A,B}(\sigma) = \frac{\mu_N(\sigma)\,\mathbb{P}_\sigma[\tau_B < \tau_A]}{\sum_{\eta \in A} \mu_N(\eta)\,\mathbb{P}_\eta[\tau_B < \tau_A]} = \frac{\mu_N(\sigma)\,e_{A,B}(\sigma)}{\text{cap}(A,B)} \qquad \forall\,\sigma \in A. \qquad (1.48)$$

Notice that $\nu_{A,B}$ is concentrated on those starting configurations $\eta \in A$ that are at the boundary of this set.

Proposition 1.12. *Let $A, B \subset \mathcal{S}_N$ with $A \cap B = \emptyset$. Then*

$$\mathbb{E}_{\nu_{A,B}}[\tau_B] = \frac{1}{\text{cap}(A,B)} \sum_{\sigma \notin B} \mu_N(\sigma)\, h_{A,B}(\sigma). \qquad (1.49)$$

Remark 1.13. We are typically interested in mean hitting times of the set B starting the process in a single configuration $\sigma \in A$. For this reason, it appears appealing to replace in (1.49) the subset A by the singleton $\{\sigma\}$. However, this results in the challenging task of computing precisely the capacity $\text{cap}(\{\sigma\},B)$ between a configuration and a set, and in most applications this cannot be achieved. On the other hand, we are able to establish sharp estimates on the right-hand side of (1.49), and hence on an averaging of mean hitting times with respect to $\nu_{A,B}$, in situations when the sets A and B are sufficiently large. In the next chapter, we derive pointwise estimates on the mean hitting times by controlling the fluctuations of $\mathbb{E}_\sigma[\tau_B]$ within A.

We present three different proofs of this proposition. While the first one, which can be found in [53], uses purely probabilistic arguments, the second proof relies on the Green function representation (1.36). It was first given in [14]. The third one is as well a rather simple proof bases on Green's second identity.

Proof of Proposition 1.12 via last exit decomposition. First, let us define the last exit time $L_{A,B}$ from A before hitting B as

$$L_{A,B} := \sup\{0 \le t < \tau_B \mid \sigma(t) \in A\}. \qquad (1.50)$$

Then we have for all $\eta \notin B$,

$$\mu_N(\eta)\, \mathbb{P}_\eta[\tau_A < \tau_B] = \mu_N(\eta)\, \mathbb{P}_\eta[L_{A,B} > 0]$$
$$= \sum_{s=1}^{\infty} \sum_{\sigma \in A} \mu_N(\eta)\, \mathbb{P}_\eta[L_{A,B} = s,\, \sigma(s) = \sigma]$$
$$= \sum_{s=1}^{\infty} \sum_{\sigma \in A} \mu_N(\eta)\, \mathbb{P}_\eta[\sigma(s) = \sigma,\, s < \tau_B]\, \mathbb{P}_\sigma[\tau_B < \tau_A], \quad (1.51)$$

where we used the Markov property in the last step. By reversibility, we have that

$$\mu_N(\eta)\, \mathbb{P}_\eta[\sigma(s) = \sigma,\, s < \tau_B] = \mu_N(\sigma)\, \mathbb{P}_\sigma[\sigma(s) = \eta,\, s < \tau_B]. \quad (1.52)$$

In view of (1.48), this implies that

$$\mathbb{E}_{\nu_{A,B}}\!\left[\sum_{s=0}^{\tau_B-1} \mathbb{1}_{\sigma(s)=\eta}\right] - \frac{\mu_N(\eta)\, \mathbb{P}_\eta[\tau_B < \tau_A]}{\mathrm{cap}(A,B)}\, \mathbb{1}_{\eta \in A} = \frac{\mu_N(\eta)\, \mathbb{P}_\eta[\tau_A < \tau_B]}{\mathrm{cap}(A,B)} \quad (1.53)$$

By summing over all configurations η outside B, we obtain (1.49). $\square$

Proof of Proposition 1.12 via Green function. For any $\sigma \in A$, multiply the representation (1.37) of the mean hitting time w_B by $\mu_N(\sigma)\, e_{A,B}(\sigma)$. Summing over all $\sigma \in A$ gives

$$\sum_{\sigma \in A} \mu_N(\sigma)\, e_{A,B}(\sigma)\, w_B(\sigma) = \sum_{\sigma \in A} \sum_{\eta \notin B} \mu_N(\sigma)\, G_B(\sigma,\eta)\, e_{A,B}(\sigma)$$
$$= \sum_{\sigma \in A} \sum_{\eta \notin B} \mu_N(\eta)\, e_{A,B}(\sigma)\, G_B(\eta,\sigma)$$
$$= \sum_{\eta \notin B} \mu_N(\eta)\, h_{A,B}(\eta), \quad (1.54)$$

where we used (1.35) in the second step and the representation (1.36) of the equilibrium potential, $h_{A,B}$, via the Green function in the third step. Normalizing the measure on the left-hand side yields the assertion. $\square$

Proof of Proposition 1.12 via Green's second identity. Applying (1.39) to the functions w_B and $h_{A,B}$ yields

$$\sum_{\sigma \in A} \mu_N(\sigma)\, e_{A,B}(\sigma)\, w_B(\sigma) = \sum_{\sigma \notin B} \mu_N(\sigma)\, h_{A,B}(\sigma). \quad (1.55)$$

Normalizing the measure on the left-hand side immediately yields the assertion. $\square$

We will now prove a splitting lemma which will be important in the investigation of the distribution of normalized hitting times.

1.5. Equilibrium potential, capacity, and mean hitting times

Lemma 1.14. Let $A, B \subset \mathcal{S}_N$ with $A \cap B = \emptyset$. Then

$$\mathbb{E}_{\nu_{A,B}}[\tau_B] = \frac{\mathbb{E}_{\mu_A}[\tau_A \mathbf{1}_{\tau_A < \tau_B}] + \mathbb{E}_{\mu_A}[\tau_B \mathbf{1}_{\tau_B < \tau_A}]}{\mathbb{P}_{\mu_A}[\tau_B < \tau_A]}. \quad (1.56)$$

We give three proofs, one purely probabilistic and two purely analytic.

Proof of Lemma 1.14 via strong Markov property. Since $1 = \mathbf{1}_{\tau_B < \tau_A} + \mathbf{1}_{\tau_A < \tau_B}$ the strong Markov property implies that

$$\begin{aligned}
\mathbb{E}_{\mu_A}[\tau_B] &= \mathbb{E}_{\mu_A}[\tau_B \mathbf{1}_{\tau_B < \tau_A}] + \mathbb{E}_{\mu_A}\left[\mathbf{1}_{\tau_A < \tau_B} \mathbb{E}_{\sigma(\tau_A)}[\tau_A + (\tau_B - \tau_A)]\right] \\
&= \mathbb{E}_{\mu_A}[\tau_B \mathbf{1}_{\tau_B < \tau_A}] + \mathbb{E}_{\mu_A}[\tau_A \mathbf{1}_{\tau_A < \tau_B}] \\
&\quad + \sum_{\eta \in A} \mathbb{P}_\sigma[\tau_A < \tau_B, \sigma(\tau_A) = \eta] \, \mathbb{E}_\eta[\tau_B].
\end{aligned}$$

As a consequence of the reversibility of the dynamics, we have that

$$\sum_{\sigma \in A} \mu_N(\sigma) \, \mathbb{P}_\sigma[\tau_A < \tau_B, \sigma(\tau_A) = \eta] = \mu_N(\eta) \, \mathbb{P}_\eta[\tau_A < \tau_B], \quad \forall \eta \in A. \quad (1.57)$$

Combining the last two equations and solving for $\mathbb{E}_\sigma[\tau_B]$ yields

$$\frac{1}{\mu_N[A]} \sum_{\eta \in A} \mu_N(\eta) \, \mathbb{P}_\eta[\tau_B < \tau_A] \, \mathbb{E}_\eta[\tau_B] = \mathbb{E}_{\mu_A}[\tau_B \mathbf{1}_{\tau_B < \tau_A}] + \mathbb{E}_{\mu_A}[\tau_A \mathbf{1}_{\tau_A < \tau_B}]$$

which, together with the definition of the last exit distribution, is equivalent to (1.56). This completes the proof. $\square$

Proof of Lemma 1.14 via Green's function. Our starting point is the representation of w_B in terms of the Green function as given in (1.37), which we rewrite as

$$w_B(\sigma) = \sum_{\eta \in A} G_B(\sigma, \eta) + \sum_{\eta \notin A \cup B} G_B(\sigma, \eta) \, (h_{A,B}(\eta) + h_{B,A}(\eta)), \quad (1.58)$$

taking advantage of the fact that $h_{A,B}(\eta) + h_{B,A}(\eta) = \mathbb{P}_\eta[\tau_A < \tau_B] + \mathbb{P}_\eta[\tau_B < \tau_A] = 1$, for all $\eta \notin A \cup B$. Multiplying both sides with $\mu_N(\sigma) e_{A,B}(\sigma)$ and summing over all $\sigma \in A$ gives

$$\sum_{\sigma \in A} \mu_N(\sigma) e_{A,B}(\sigma) w_B(\sigma)$$

$$= \mu_N[A] + \sum_{\substack{\sigma \in A \\ \eta \notin A \cup B}} e_{A,B}(\sigma) \mu_N(\eta) G_B(\eta, \sigma) \, (h_{A,B}(\eta) + h_{B,A}(\eta)), \quad (1.59)$$

where we used (1.35) as well as the representation, (1.36), of the harmonic function $h_{A,B}$ via the Green's function. Exploiting that, for all $\sigma \notin A \cup B$, the function $w_{A,B}$ solves the Poisson equation $(L_N w_{A,B})(\sigma) = -h_{A,B}(\sigma)$ subject to zero boundary condition of $A \cup B$, we can replace the harmonic function $h_{A,B}$ in (1.59) by the left-hand side of the Poisson equation. By applying Green's second identity to the functions $w_{A,B}$ and $G_{B^c}(\,\cdot\,,\sigma)$ with $\sigma \in A$ we have

$$\sum_{\eta \notin A \cup B} \mu_N(\eta)\, G_B(\eta,\sigma)\, (L_N w_{A,B})(\eta) = -\sum_{\eta \in A} \mu_N(\eta)\, G_B(\eta,\sigma)\, (L_N w_{A,B})(\eta), \tag{1.60}$$

whereas the same holds true when $w_{A,B}$ is replaced by $w_{B,A}$. Carrying out the summation over $\sigma \in A$ and taking into account (1.36), yields

$$\sum_{\sigma \in A} \mu_N(\sigma)\, e_{A,B}(\sigma)\, w_B(\sigma) = \sum_{\eta \in A} \mu_N(\sigma) \left(1 + (L_N w_{A,B})(\eta) + (L_N w_{B,A})(\eta) \right) \tag{1.61}$$

$$= \sum_{\eta \in A} \mu_N(\eta) \left(\mathbb{E}_\eta \big[\tau_A\, \mathbf{1}_{\tau_A < \tau_B} \big] + \mathbb{E}_\eta \big[\tau_B\, \mathbf{1}_{\tau_B < \tau_A} \big] \right).$$

The last step relies on the probabilistic interpretation of the involved analytic quantities and the boundary of A. Namely, for all $\eta \in A$ it holds that $\mathbb{E}_\eta[\tau_B\, \mathbf{1}_{\tau_B < \tau_A}] = e_{A,B}(\eta) + (L_N w_{B,A})(\eta)$ and $\mathbb{E}_\eta[\tau_A\, \mathbf{1}_{\tau_A < \tau_B}] = 1 - e_{A,B}(\eta) + (L_N w_{A,B})(\eta)$. Normalizing the measure on the left-hand side completes the proof. $\square$

Proof of Lemma 1.14 via second Green's formula. By applying Green's second identity to the functions $w_{A,B}$ and $h_{A,B}$ gives

$$\sum_{\sigma \in A} \mu_N(\sigma)\, (L_N w_{A,B})(\sigma) = \sum_{\sigma \notin A \cup B} \mu_N(\sigma)\, h_{A,B}^2(\sigma), \tag{1.62}$$

whereas its application to the functions $w_{B,A}$ and $h_{A,B}$ leads to

$$\sum_{\sigma \in A} \mu_N(\sigma)\, (L_N w_{B,A})(\sigma) = \sum_{\sigma \notin A \cup B} \mu_N(\sigma)\, h_{A,B}(\sigma)\, h_{B,A}(\sigma). \tag{1.63}$$

Adding up the last two equations, yields

$$\sum_{\sigma \in A} \mu_N(\sigma) \left((L_N w_{A,B})(\sigma) + (L_N w_{B,A})(\sigma) \right) = \sum_{\sigma \notin A \cup B} \mu_N(\sigma)\, h_{A,B}(\sigma). \tag{1.64}$$

By comparing this equation with (1.55), we immediately recover (1.61) and the remaining part of the proof follows the arguments given in the previous proof. $\square$

1.6. The Dirichlet and Thomson principle: a dual variational principle

The power of Proposition (1.12) is that it represents the mean hitting time in terms of a capacity that is a positive-definite quadratic form, as it can be seen from (1.47). Whenever a object of interests can be expressed as a positive-definite quadratic form, it is often possible to establish dual variational principles from which upper and lower bound for it can be constructed. Notice that many variational methods are based on

1.6. The Dirichlet and Thomson principle: a dual variational principle

this tool. Our presentation follows in parts the one given by Lieberstein [64] in the context of partial differential equations.

The bilinear form $\mathcal{E}(g,h)$ associated to the Dirichlet form $\mathcal{E}(g)$ is given by

$$\mathcal{E}(g,h) = \frac{1}{2} \sum_{\sigma,\eta \in \mathcal{S}_N} \mu_N(\sigma) p_N(\sigma,\eta) \left(g(\sigma) - g(\eta)\right) \left(h(\sigma) - h(\eta)\right). \quad (1.65)$$

An immediate consequence of the Cauchy-Schwarz inequality is that

$$\mathcal{E}(g,h)^2 \leq \mathcal{E}(g)\,\mathcal{E}(h). \quad (1.66)$$

Recall that the equilibrium potential, $h_{A,B}$, is defined as the solution of the Laplace equation

$$\begin{cases} (L_N f)(\sigma) = 0, & \sigma \in (A \cup B)^c \\ f(\sigma) = \mathbb{1}_A(\sigma), & \sigma \in A \cup B. \end{cases} \quad (1.67)$$

If a function $h \in L^2(\mathcal{S}_N, \mu_N)$ has the property that $h = \mathbb{1}_A$ on $A \cup B$, then by Green's first identity

$$\mathcal{E}(h_{A,B}, h) = -\sum_{\eta \in \mathcal{S}_N} \mu_N(\eta) h(\eta) \left(L_N h_{A,B}\right)(\eta) = \sum_{\eta \in A} \mu_N(\eta) e_{A,B}(\eta) \quad (1.68)$$

In view of (1.44) and (1.47), for such a function h, we have that $\mathrm{cap}(A,B) \leq \mathcal{E}(h)$ by (1.66). Since equality is obtains, if we choose for h the equilibrium potential, we have proven the well known

Proposition 1.15 (Dirichlet principle). *For any non-empty disjoint sets $A, B \subset \mathcal{S}_N$,*

$$\mathrm{cap}(A,B) = \min_{h \in \mathcal{H}_{A,B}} \mathcal{E}(h) \quad (1.69)$$

where $\mathcal{H}_{A,B} := \{h \colon \mathcal{S}_N \to [0,1] \mid h|_A \equiv 1,\ h|_B \equiv 0\}$.

Remark 1.16. The importance of the Dirichlet principle is that it yields computable upper bounds for capacities by choosing suitable test functions h. As we will see in chapter 5, with proper physical insight, it is often possible to guess a reasonable test function.

Remark 1.17. An immediate corollary of the Dirichlet principle is *Rayleigh's monotonicity law*. It allows to derive a lower bound for capacities by using the monotonicity of the Dirichlet form in the transition probabilities to compare the original process with a simplified one. This is well known in the language of electrical networks, see [44].

Instead of considering a function h which coincides with $h_{A,B}$ on $A \cup B$, suppose that $h \in L^2(\mathcal{S}_N, \mu_N)$ is super-harmonic on B^c, i.e. $(L_N h)(\sigma) \leq 0$ for all $\sigma \in B^c$. Then by Green's first identity

$$\mathcal{E}(h_{A,B}, h) = -\sum_{\eta \in \mathcal{S}_N} \mu_N(\sigma) h_{A,B}(\sigma) (L_N h)(\sigma) \geq -\sum_{\eta \in A} \mu_N(\sigma) (L_N h)(\sigma), \tag{1.70}$$

where we used in the second step that on $(A \cup B)^c$ the equilibrium potential is non-negative. Hence, from (1.66), we get that

$$\operatorname{cap}(A, B) \geq \frac{\left(\sum_{\eta \in A} \mu_N(\eta) (L_N h)(\eta)\right)^2}{\mathcal{E}(h)} \tag{1.71}$$

where equality is obtain for $h = h_{A,B}$. Further, notice that the value of the right-hand side of (1.71) is invariant with respect to multiplication of h by a constant. This allows to choose h is such a way that the numerator of (1.71) is equal to 1. Thus, we have proven a version of Sir William Thomson's (Lord Kelvin's) principle

Proposition 1.18. *For any non-empty disjoint sets $A, B \subset \mathcal{S}_N$,*

$$\operatorname{cap}(A, B) = \max_{h \in \mathfrak{H}_{A,B}} \frac{\left(\sum_{\eta \in A} \mu_N(\eta) (L_N h)(\eta)\right)^2}{\mathcal{E}(h)}. \tag{1.72}$$

where $\mathfrak{H}_{A,B} := \{h : \mathcal{S}_N \to [0, 1] \mid (L_N h)(\eta) \leq 0, \forall \eta \notin B\}$.

The Thomson principle is well known in the context of random walks and electrical networks. However, in that language it is differently formulated in terms of unit flows, see [44, 83]. For this reason, we will briefly discuss the connection between these formulations.

Definition 1.19. *Given two non-empty disjoint sets $A, B \subset \mathcal{S}_N$, and a map $f : \mathcal{S}_N \times \mathcal{S}_N \to \mathbb{R}$ such that, for all $\sigma, \eta \in \mathcal{S}_N$, $f(\sigma, \eta) = -f(\eta, \sigma)$ and $p_N(\sigma, \eta) = 0$ implies $f(\sigma, \eta) = 0$.*

(a) *The map f is a flow from A to B, if it satisfies Kirchhoff's law in $(A \cup B)^c$, that is*

$$\sum_{\eta \in \mathcal{S}_N} f(\sigma, \eta) = \sum_{\eta \in \mathcal{S}_N} f(\eta, \sigma), \qquad \forall \sigma \in (A \cup B)^c. \tag{1.73}$$

(b) *The map f is a unit flow from A to B, if it is a flow and the strength of f is equal to 1, that is*

$$\sum_{\sigma \in A} \sum_{\eta \in \mathcal{S}_N} f(\sigma, \eta) = 1 = \sum_{\eta \in \mathcal{S}_N} \sum_{\sigma \in B} f(\eta, \sigma). \tag{1.74}$$

With this definition of flows on the network given by $(p(\sigma,\eta))$ in mind, let us rewrite the bilinear form $\mathcal{E}(g,h)$ as

$$\mathcal{E}(g,h) = \frac{1}{2} \sum_{\sigma,\eta \in \mathcal{S}_N} \frac{1}{\mu_N(\sigma) p_N(\sigma,\eta)} (\mu_N p_N \nabla g)(\sigma,\eta)(\mu_N p_N \nabla h)(\sigma,\eta), \quad (1.75)$$

where we introduced the notation $(\mu_N p_N \nabla g)(\sigma,\eta) := \mu_N(\sigma) p_N(\sigma,\eta)(g(\sigma) - g(\eta))$. Now observe that if the function g is harmonic then $\mu_N p_N \nabla g$ is a flow. Thus, in view of (1.75), given two flows f, ϕ we define a bilinear form through

$$\mathcal{D}(f,\phi) := \frac{1}{2} \sum_{\sigma,\eta \in \mathcal{S}_N} \frac{1}{\mu_N(\sigma) p_N(\sigma,\eta)} f(\sigma,\eta) \phi(\sigma,\eta), \quad \mathcal{D}(f) := \mathcal{D}(f,f). \quad (1.76)$$

In particular, $\mathcal{D}(\mu_N p_N \nabla h_{A,B}) = \mathrm{cap}(A,B)$. Since for any unit flow f,

$$\mathcal{D}(\mu_N p_N \nabla h_{A,B}, f) = \frac{1}{2} \sum_{\sigma,\eta \in \mathcal{S}_N} (h_{A,B}(\sigma) - h_{A,B}(\eta)) f(\sigma,\eta)$$

$$= \sum_{\sigma \in A} \sum_{\eta \in \mathcal{S}_N} f(\sigma,\eta) = 1 \quad (1.77)$$

an application of Cauchy-Schwarz's inequality reveals $\mathrm{cap}(A,B) \geq 1/\mathcal{D}(f)$, whereas equality holds for the unit flow which is given in terms of the harmonic function. Thus, we have proven

Proposition 1.20 (Thomson principle). *For any non-empty disjoint sets $A, B \subset \mathcal{S}_N$,*

$$\mathrm{cap}(A,B) = \max_{f \in \mathfrak{U}_{A,B}} \frac{1}{\mathcal{D}(f)}, \quad (1.78)$$

where $\mathfrak{U}_{A,B}$ denotes the space of unit flows from A to B.

1.7. The Berman-Konsowa principle

The variational principles, we have presented in the previous section, rely on the application of the Cauchy-Schwarz inequality to the bilinear form $\mathcal{E}(g,h)$ and $\mathcal{D}(f,\phi)$, respectively. We will now describe a little-known variational principles for capacities that does not use (1.66). It was first proven by Berman and Konsowa in [4]. Our presentation will follow the arguments first given in [6] and then in [14].

Definition 1.21. Given two non-empty disjoint sets $A, B \subset \mathcal{S}_N$, a map $f: \mathcal{S}_N \times \mathcal{S}_N \to [0,\infty)$ such that $p_N(\sigma,\eta) = 0$ implies $f(\sigma,\eta) = 0$ for all $\sigma,\eta \in \mathcal{S}_N$. Then f is a loop-free non-negative unit flow, if it satisfies the following properties

  (i) if $f(\sigma,\eta) > 0$ then $f(\eta,\sigma) = 0$,
  (ii) f satisfies Kirchhoff's law in $(A \cup B)^c$,
  (iii) the strength of f is equal to 1,
  (iv) any path, γ, from A to B such that $f(\sigma,\eta) > 0$ for all $(\sigma,\eta) \in \gamma$ is self-avoiding.

In what follows, our objective is to show that each loop-free non-negative unit flow f give rise to a probability measure $\mathbb{P}^f$ on self-avoiding paths. For a given f, we define $\mathcal{F}(\sigma) := \sum_{\eta \in \mathcal{S}_N} f(\sigma, \eta)$ which is positive for all $\sigma \notin B$. Then $\mathbb{P}^f$ is the law of a Markov chain $\{\xi(t)\}$ that is stopped at the arrival of B with initial distribution $\mathbb{P}^f[\xi(0) = \sigma] = \mathcal{F}(\sigma)\mathbb{1}_A(\sigma)$ and transition probabilities

$$q^f(\sigma, \eta) = \frac{f(\sigma, \eta)}{\mathcal{F}(\eta)}, \qquad \sigma \notin B. \tag{1.79}$$

Hence, for a path $\gamma = (\sigma^0, \ldots, \sigma^r)$ with $\sigma^0 \in A$, $\sigma^r \in B$ and $\sigma^k \in \mathcal{S}_N \setminus (A \cup B)$, $\forall k = 1, \ldots r - 1$ we have

$$\mathbb{P}^f[\xi = \gamma] = \mathcal{F}(\sigma^0) \prod_{k=0}^{r-1} \frac{f(\sigma^k, \sigma^{k+1})}{\mathcal{F}(\sigma^k)} \tag{1.80}$$

where we used the convention that $0/0 = 0$. By using Kirchhoff's law, we obtain the following representation for the probability that $\{\xi(t)\}$ passes through an edge (σ, η)

$$\mathbb{P}^f[\xi \ni (\sigma, \eta)] = \sum_{\gamma} \mathbb{P}^f[\xi = \gamma] \mathbb{1}_{(\sigma, \eta) \in \gamma} = f(\sigma, \eta). \tag{1.81}$$

The equation (1.81) gives rise to the following *partition of unity*

$$\mathbb{1}_{f(\sigma,\eta)>0} = \sum_{\gamma} \mathbb{P}^f[\xi = \gamma] \frac{1}{f(\sigma, \eta)} \mathbb{1}_{(\sigma,\eta) \in \gamma} \tag{1.82}$$

Employing this representation, for every $h \in L^2(\mathcal{S}_N, \mu_N)$ which satisfy the boundary condition $h = \mathbb{1}_A$ on $A \cup B$, we can bound the Dirichlet form from below by

$$\mathcal{E}(h) \geq \sum_{(\sigma, \eta)} \mu_N(\sigma) p_N(\sigma, \eta) (h(\sigma) - h(\eta))^2 \mathbb{1}_{f(\sigma,\eta)>0}$$

$$= \sum_{\gamma} \mathbb{P}^f[\xi = \gamma] \sum_{(\sigma,\eta) \in \gamma} \frac{\mu_N(\sigma) p_N(\sigma, \eta)}{f(\sigma, \eta)} (h(\sigma) - h(\eta))^2 \tag{1.83}$$

Let us now minimize the expressions above over all function h with $h|_A = 1$ and $h|_B = 0$. By interchanging the minimum and the sum on the right-hand side of (1.83), we are left with the task to solve optimization problems along one-dimensional path from A to B whose minimizer are explicitly known [11]. As a result, we get

$$\mathrm{cap}(A, B) \geq \sum_{\gamma} \mathbb{P}^f[\xi = \gamma] \left(\sum_{(\sigma,\eta) \in \gamma} \frac{f(\sigma, \eta)}{\mu_N(\sigma) p_N(\sigma, \eta)} \right)^{-1}. \tag{1.84}$$

An important loop-free unit flow is the *harmonic flow*, $f_{A,B}$, which is defined in terms of the equilibrium potential

$$f_{A,B}(\sigma, \eta) := \frac{\mu_N(\sigma) p_N(\sigma, \eta)}{\mathrm{cap}(A, B)} \left[h_{A,B}(\sigma) - h_{A,B}(\eta) \right]_+. \tag{1.85}$$

1.8. Renewal equation

It is easy to verify that $f_{A,B}$ satisfies the conditions (i)–(iv). While, (i) is obvious, (ii) is a consequence that $h_{A,B}$ is harmonic in $(A \cup B)^c$ and (iii) follows from (1.44). Condition (iv) comes from the fact that the harmonic flow only moves in directions where $h_{A,B}$ decreases. Since for $f_{A,B}$ equality is obtained in (1.84) $\mathbb{P}^f$-a.s., this proves

Proposition 1.22. *(Berman-Konsowa principle)* Let $A, B \subset S_N$ be disjoint. Then, with the notations introduced above,

$$\operatorname{cap}(A, B) = \max_{f \in \mathcal{U}_{A,B}} \mathbb{E}^f \left[\left(\sum_{(\sigma,\eta) \in \gamma} \frac{f(\sigma, \eta)}{\mu_N(\sigma) p_N(\sigma, \eta)} \right)^{-1} \right], \tag{1.86}$$

where $\mathcal{U}_{A,B}$ denotes the space of all loop-free non-negative unit flows from A to B.

The nice feature of this variational principle is that any flow gives rise to a computable lower bound. In order to obtain upper and lower bounds on the capacity that differs only a little, the strategy is to find first a good approximation of the harmonic function, $h_{A,B}$, by a test function h and to construct, in a second step, a test flow f from it. By plugging f into (1.84), the quality of the approximation can be controlled.

Remark 1.23. Given a loop-free non-negative unit flow f, then, by Jensen's inequality

$$\mathbb{E}^f \left[\left(\sum_{(\sigma,\eta) \in \gamma} \frac{f(\sigma, \eta)}{\mu_N(\sigma) p_N(\sigma, \eta)} \right)^{-1} \right] \geq \left(\mathbb{E}^f \left[\sum_{(\sigma,\eta) \in \gamma} \frac{f(\sigma, \eta)}{\mu_N(\sigma) p_N(\sigma, \eta)} \right] \right)^{-1} = \frac{1}{\mathcal{D}(f)}. \tag{1.87}$$

Since the Thomson principle also holds for loop-free non-negative unit flows, this shows that the lower bound on the capacity obtained by the Berman-Konsowa principle provides, in principle, a better approximation compared to the Thomson principle.

1.8. Renewal equation

In earlier works, see for example [15, 16], the following renewal equations played a crucial role in order to estimate harmonic functions or to control the Laplace transforms of hitting times. Let $\sigma, \eta \in S_N$, $\eta \notin B \subset S_N$ and $u : \mathbb{R} \to \mathbb{R}$, then

$$\mathbb{P}_\sigma[\tau_\eta < \tau_B] = \frac{\mathbb{P}_\sigma[\tau_\eta < \tau_{B \cup \sigma}]}{\mathbb{P}_\sigma[\tau_{B \cup \eta} < \tau_\sigma]} \quad \text{and} \quad \mathbb{E}_\sigma\left[e^{-u(\lambda)\tau_B}\right] = \frac{\mathbb{E}_\sigma\left[e^{-u(\lambda)\tau_B} \mathbb{1}_{\tau_B < \tau_\sigma}\right]}{1 - \mathbb{E}_\sigma\left[e^{-u(\lambda)\tau_\sigma} \mathbb{1}_{\tau_\sigma < \tau_B}\right]} \tag{1.88}$$

where the second equation holds for all λ for which the left-hand side exists. However, as it was pointed out in [6, 5, 14], pointwise renewal equation are of limited use in the general context. Instead of studying the process in a given point σ, we show how an averaged version of the standard renewal argument can be established.

Lemma 1.24. Let $A, B, X \subset \mathcal{S}_N$ be mutually disjoint. Then

$$\mathbb{P}_{\nu_{X,A\cup B}}[\tau_A < \tau_B] = \frac{\mathbb{P}_{\mu_X}[\tau_A < \tau_{B\cup X}]}{\mathbb{P}_{\mu_X}[\tau_{A\cup B} < \tau_X]} \leq \min\left\{\frac{\operatorname{cap}(X, A)}{\operatorname{cap}(X, B)}, 1\right\}. \qquad (1.89)$$

We give three proofs, one purely probabilistic using strong Markov property and two purely analytic ones using the Green's function and the second Green's identity.

Proof of Lemma 1.24 via strong Markov property. By starting the process in the reversible measure μ_X, we obtain

$$\mathbb{P}_{\mu_X}[\tau_A < \tau_B] = \mathbb{P}_{\mu_X}[\tau_A < \tau_{B\cup X}] + \sum_{\eta \in X} \mathbb{P}_{\mu_X}[\tau_X < \tau_{A\cup B}, \sigma(\tau_X) = \eta]\, \mathbb{P}_\eta[\tau_A < \tau_B]$$

$$(1.90)$$

As a consequence of the reversibility, we have that

$$\sum_{\sigma \in X} \mu_N(\sigma)\, \mathbb{P}_\sigma[\tau_X < \tau_{A\cup B}, \sigma(\tau_X) = \eta] = \mu_N(\eta)\, \mathbb{P}_\eta[\tau_X < \tau_{A\cup B}]. \qquad (1.91)$$

Hence, this allows us to rewrite (1.90) as

$$\frac{1}{\mu_N[X]} \sum_{\sigma \in X} \mu_N(\sigma)\, \mathbb{P}_\eta[\tau_{A\cup B} < \tau_X]\, \mathbb{P}_\eta[\tau_A < \tau_B] = \mathbb{P}_{\mu_X}[\tau_A < \tau_{B\cup X}]. \qquad (1.92)$$

Normalizing the measure on the left-hand side (1.92) and comparing the result with (1.48), yields

$$\mathbb{P}_{\nu_{X,A\cup B}}[\tau_A < \tau_B] = \frac{\mathbb{P}_{\mu_X}[\tau_A < \tau_{B\cup X}]}{\mathbb{P}_{\mu_X}[\tau_{A\cup B} < \tau_X]} \leq \frac{\mathbb{P}_{\mu_X}[\tau_A < \tau_X]}{\mathbb{P}_{\mu_X}[\tau_B < \tau_X]} = \frac{\operatorname{cap}(X, A)}{\operatorname{cap}(X, B)},$$

where we used elementary monotonicity properties and the definition of capacities. Obviously, this bound is only useful if $\operatorname{cap}(X, A)/\operatorname{cap}(X, B) < 1$. $\square$

Proof of Lemma 1.24 via Green's function. To start with, consider the so called Poisson kernel representation of $h_{A,B}$, as defined in (1.43), which reads

$$h_{A,B}(\sigma) = \sum_{\eta \in A \cup B} \frac{\mu_N(\eta)}{\mu_N(\sigma)} h_{A,B}(\eta)\, (L_N G_{A\cup B})(\eta, \sigma). \qquad (1.93)$$

In particular, for any $\sigma \in X$, we get that

$$h_{A,B}(\sigma) = \sum_{\eta \in A \cup B} \frac{\mu_N(\eta)}{\mu_N(\sigma)} h_{A,B\cup X}(\eta)\, (L_N G_{A\cup B})(\eta, \sigma). \qquad (1.94)$$

By applying the second Green's identity to the functions $h_{A,B\cup X}$ and $G_{A\cup B}(\cdot, \sigma)$, we obtain

$$\sum_{\eta \in A\cup B} \mu_N(\eta)\, h_{A,B\cup X}(\eta)\, (L_N G_{A\cup B})(\eta, \sigma) = \sum_{\eta \in X} \mu_N(\eta)\, e_{B\cup X, A}(\eta)\, G_{A\cup B}(\eta, \sigma).$$

$$(1.95)$$

1.8. Renewal equation

Hence, combining (1.94) and (1.95) yields

$$h_{A,B}(\sigma) = \sum_{\eta \in X} \frac{\mu_N(\eta)}{\mu_N(\sigma)} e_{B \cup X, A}(\eta) G_{A \cup B}(\eta, \sigma). \qquad (1.96)$$

Thus, if we multiply both sides with $\mu_N(\sigma) e_{X, A \cup B}(\sigma)$ and sum over all $\sigma \in X$, we get

$$\sum_{\sigma \in X} \mu_N(\sigma) e_{X, A \cup B}(\sigma) h_{A,B}(\sigma) = \sum_{\sigma, \eta \in X} \mu_N(\eta) e_{B \cup X, A}(\eta) G_{A \cup B}(\eta, \sigma) e_{X, A \cup B}(\sigma)$$

$$= \sum_{\eta \in X} \mu_N(\eta) e_{B \cup X, A}(\eta), \qquad (1.97)$$

where we used in the second step the representation (1.36) of $h_{X, A \cup B}$ in terms of the Green function and that $h_{X, A \cup B}(\eta) = 1$, for all $\eta \in X$. Dividing both sides of (1.97) by $\mathrm{cap}(X, A \cup B)$ and using the definition of the last exit biased distribution (1.48), we obtain

$$\mathbb{P}_{\nu_{X, A \cup B}}[\tau_A < \tau_B] = \frac{\sum_{\eta \in X} \mu_N(\eta) e_{B \cup X, A}}{\mathrm{cap}(X, A \cup B)} = \frac{\mathbb{P}_{\mu_X}[\tau_A < \tau_{B \cup X}]}{\mathbb{P}_{\mu_X}[\tau_{A \cup B} < \tau_X]} \leq \frac{\mathrm{cap}(X, A)}{\mathrm{cap}(X, B)}.$$

This concludes the proof. □

Proof of Lemma 1.24 via second Green's identity. By applying the second Green's identity to the functions $h_{A,B}$ and $h_{X, A \cup B}$, we get

$$-\sum_{\sigma \in X} \mu_N(\sigma) \left(L_N h_{X, A \cup B} \right)(\sigma) h_{A,B}(\sigma) = \sum_{\sigma \in A} \mu_N(\sigma) \left(L_N h_{X, A \cup B} \right)(\sigma). \qquad (1.98)$$

An application of the second Green's identity to the functions $h_{X, A \cup B}$ and $h_{A, B \cup X}$ reveals that

$$\sum_{\sigma \in A} \mu_N(\sigma) \left(L_N h_{X, A \cup B} \right)(\sigma) = \sum_{\sigma \in X} \mu_N(\sigma) \left(L_N h_{A, B \cup X} \right)(\sigma). \qquad (1.99)$$

Hence, by combing both equations, we obtain

$$\sum_{\sigma \in X} \mu_N(\sigma) e_{X, A \cup B}(\sigma) h_{A,B}(\sigma) = \sum_{\sigma \in X} \mu_N(\sigma) e_{B \cup X, A}(\sigma). \qquad (1.100)$$

By dividing both sides of (1.100) by $\mathrm{cap}(X, A \cup B)$ and using the definition of the last exit biased distribution as well as the probabilistic interpretation of the equilibrium measures $e_{B \cup X, A}(\eta)$ and $e_{A \cup B, X}(\eta)$, allows to deduce (1.89). □

Our next goal is to show an averaged version of the renewal equation for the Laplace transform. In the previous sections we already presented the profound connection between potential theory and the Laplace transform. Given two disjoint subsets $A, B \subset$

S_N, our starting point is a specific probability measure ρ_λ on A which is defined for any $\lambda \geq 0$ by

$$\rho_\lambda(\sigma) = \frac{\mu_N(\sigma)\left(1 - \mathbb{E}_\sigma\left[e^{-\lambda \tau_A} 1_{\tau_A < \tau_B}\right]\right)}{\sum_{\eta \in A} \mu_N(\eta)\left(1 - \mathbb{E}_\eta\left[e^{-\lambda \tau_A} 1_{\tau_A < \tau_B}\right]\right)} = \frac{\mu_N(\sigma)\, e^\lambda_{A,B}(\sigma)}{\sum_{\eta \in A} \mu_N(\eta)\, e^\lambda_{A,B}(\eta)}. \quad (1.101)$$

Let us point out that ρ_λ is a generalization of the last exist biased distribution which can be retrieved from ρ_λ by setting $\lambda = 0$.

The following renewal equation will be used in Section 3.4.

Lemma 1.25. *Let $A, B \subset S_N$ with $A \cap B \neq \emptyset$ and ρ_λ as defined above. Then*

$$\mathbb{E}_{\rho_\lambda}\left[e^{-\lambda \tau_B}\right] = \frac{\mathbb{E}_{\mu_A}\left[e^{-\lambda \tau_B} 1_{\tau_B < \tau_A}\right]}{1 - \mathbb{E}_{\mu_A}\left[e^{-\lambda \tau_A} 1_{\tau_A < \tau_B}\right]}. \quad (1.102)$$

Again, we give three proofs, one purely probabilistic using the strong Markov property and two purely analytic using the Green's function and the second Green's identity, respectively.

Proof of Lemma 1.25 via strong Markov property. Since $1 = 1_{\tau_B < \tau_A} + 1_{\tau_A < \tau_B}$, the strong Markov property yields

$$\mathbb{E}_{\mu_A}\left[e^{-\lambda \tau_B}\right] = \mathbb{E}_{\mu_A}\left[e^{-\lambda \tau_B} 1_{\tau_B < \tau_A}\right] \\ + \sum_{\eta \in A} \mathbb{E}_{\mu_A}\left[e^{-\lambda \tau_A} 1_{\tau_A < \tau_B} 1_{\sigma(\tau_A) = \eta}\right] \mathbb{E}_\eta\left[e^{-\lambda \tau_B}\right]. \quad (1.103)$$

Let us now consider the last term. By reversibility, it holds that

$$\sum_{\sigma \in A} \mu_N(\sigma)\, \mathbb{E}_\sigma\left[e^{-\lambda \tau_A} 1_{\tau_A < \tau_B} 1_{X(\tau_A) = \eta}\right] = \mu_N(\eta)\, \mathbb{E}_\eta\left[e^{-\lambda \tau_A} 1_{\tau_A < \tau_B}\right]. \quad (1.104)$$

Combining (1.104) with (1.103) and solving for $\mathbb{E}_\eta\left[e^{-\lambda \tau_B}\right]$ yields

$$\frac{1}{\mu_N[A]} \sum_{\eta \in A} \mu_N(\sigma)\left(1 - \mathbb{E}_\eta\left[e^{-\lambda \tau_A} 1_{\tau_A < \tau_B}\right]\right) \mathbb{E}_\eta\left[e^{-\lambda \tau_B}\right] = \mathbb{E}_{\mu_A}\left[e^{-\lambda \tau_B} 1_{\tau_B < \tau_A}\right].$$

Since $\lambda \geq 0$, we can normalize the measure on the left-hand side. In view of (1.101), the normalization constant is equal to $\mu_N[A] - \mu_N[A]\, \mathbb{E}_{\mu_A}\left[e^{-\lambda \tau_A} 1_{\tau_A < \tau_B}\right]$ and hence, (1.102) follows. $\square$

Proof of Lemma 1.25 via Green's function. For any $\sigma \in A$ consider the Poisson kernel representation of w_B^λ given by

$$w_B^\lambda(\sigma) = \sum_{\eta \in B} \frac{\mu_N(\eta)}{\mu_N(\sigma)} (L_N^\lambda G_B^\lambda)(\eta, \sigma) = \sum_{\eta \in B} \frac{\mu_N(\eta)}{\mu_N(\sigma)} (L_N^\lambda G_B^\lambda)(\eta, \sigma)\, h_{B,A}^\lambda(\eta).$$

$$(1.105)$$

Applying the second Green's identity to the functions $h_{B,A}^\lambda$ and $G_B^\lambda(\,\cdot\,,\sigma)$, we obtain

$$\sum_{\eta \in B} \mu_N(\eta)\, h_{B,A}^\lambda(\eta) \left(L_N^\lambda G_B^\lambda\right)(\eta,\sigma) = \sum_{\eta \notin B} \mu_N(\eta)\, G_B^\lambda(\eta,\sigma) \left(L_N^\lambda h_{B,A}^\lambda\right)(\eta). \quad (1.106)$$

Hence, combining (1.105) and (1.105) and exploiting that $L_N^\lambda h_{B,A}^\lambda$ vanishes on $(A \cup B)^c$, we get

$$w_B^\lambda(\sigma) = \sum_{\eta \in A} \frac{\mu_N(\eta)}{\mu_N(\sigma)}\, G_B^\lambda(\eta,\sigma) \left(L^\lambda h_{B,A}^\lambda\right)(\eta). \quad (1.107)$$

Multiplying both sides with $\mu_N(\sigma)\, e_{A,B}^\lambda(\sigma)$ and summing over all $\sigma \in A$, provides

$$\sum_{\sigma \in A} \mu_N(\sigma)\, e_{A,B}^\lambda(\sigma)\, w_B^\lambda(\sigma) = \sum_{\eta,\sigma \in A} \mu_N(\eta) \left(L^\lambda h_{B,A}^\lambda\right)(\eta)\, G_B^\lambda(\eta,\sigma)\, e_{A,B}^\lambda(\sigma)$$

$$= \sum_{\eta \in A} \mu_N(\eta) \left(L^\lambda h_{B,A}^\lambda\right)(\eta), \quad (1.108)$$

where we used in the second step the representation (1.36) of $h_{A,B}^\lambda$ in terms of the Green function and that $h_{A,B}^\lambda(\eta) = 1$, for all $\eta \in A$. By normalizing the measure on the left-hand side and using the probabilistic interpretation (1.21) of $e_{A,B}^\lambda$ concludes the proof. □

Proof of Lemma 1.25 via second Green's formula. By applying the second Green's identity to the functions $h_{A,B}^\lambda$ and w_B^λ, we get

$$\sum_{\sigma \in A} \mu_N(\sigma) \left(L^\lambda h_{A,B}^\lambda\right)(\sigma)\, w_B^\lambda(\sigma) = -\sum_{\sigma \in B} \mu_N(\sigma) \left(L^\lambda h_{A,B}^\lambda\right)(\sigma) \quad (1.109)$$

A further application of the second Green's identity to the functions $h_{A,B}^\lambda$ and $h_{B,A}^\lambda$ yields

$$\sum_{\sigma \in A} \mu_N(\sigma) \left(L^\lambda h_{B,A}^\lambda\right)(\sigma) = \sum_{\sigma \in B} \mu_N(\sigma) \left(L^\lambda h_{A,B}^\lambda\right)(\sigma). \quad (1.110)$$

Hence, by combining these equations, we obtain

$$\sum_{\sigma \in A} \mu_N(\sigma)\, e_{A,B}^\lambda(\sigma)\, w_B^\lambda(\sigma) = -\sum_{\sigma \in A} \mu_N(\sigma)\, e_{B,A}^\lambda(\sigma). \quad (1.111)$$

Finally, dividing both sides by $\sum_{\sigma \in A} \mu_N(\sigma)\, e_{A,B}^\lambda(\sigma)$ and exploiting (1.21) as well as the definition of ρ_λ we conclude the proof. □

1.9. Almost ultrametricity of capacities

In order to derive general results under the definition of metastability, the almost ultrametricity of capacities plays a crucial role. This has been noted first in [16]. The proof given there is purely probabilistic and relies on splitting and renewal ideas. In the following, we will present two different proofs of a generalized statement which,

28   1. Basic ingredients of the potential theoretic approach

in addition, allows to control capacities between different sets. While our first proof, based heavily on analytic arguments, is maybe less intuitive from a probabilistic point of view, it is considerably simpler than the second one which takes advantage of the strong Markov property. The control obtained is crucial for the investigation in the next sections.

Lemma 1.26. Let $X, Y, A \subset S_N$ be mutually disjoint. Assume that there exists $0 < \delta < \frac{1}{C}$, where

$$C := \max\left\{\frac{\mathbb{P}_{\nu_{X,A}}[\tau_A < \tau_Y]}{\mathbb{P}_{\nu_{X,A \cup Y}}[\tau_A < \tau_Y]}, \frac{\mathbb{P}_{\nu_{Y,A}}[\tau_A < \tau_X]}{\mathbb{P}_{\nu_{Y,X}}[\tau_A < \tau_X]}\right\} = \max\{C_1, C_2\} \quad (1.112)$$

such that $\mathrm{cap}(X, A) \leq \delta\, \mathrm{cap}(X, Y)$. Then

$$1 - C_2 \delta \leq \frac{\mathrm{cap}(X, A)}{\mathrm{cap}(Y, A)} \leq \frac{1}{1 - C_1 \delta}. \quad (1.113)$$

Proof of Lemma 1.26 via analytic arguments. An application of the second Green's identity to the functions $h_{X,A}$ and $h_{Y,A}$ reveals

$$\sum_{\sigma \in X} \mu_N(\sigma)\, (Lh_{X,A})(\sigma)\, h_{Y,A}(\sigma) = \sum_{\sigma \in Y} \mu_N(\sigma)\, (Lh_{Y,A})(\sigma)\, h_{X,A}(\sigma). \quad (1.114)$$

Hence, by normalizing the measure on the left-hand side as well as the one on the right-hand side, we immediately obtain

$$1 - \mathbb{P}_{\nu_{Y,A}}[\tau_A < \tau_X] \leq \frac{\mathrm{cap}(X, A)}{\mathrm{cap}(Y, A)} = \frac{\mathbb{P}_{\nu_{Y,A}}[\tau_X < \tau_A]}{\mathbb{P}_{\nu_{X,A}}[\tau_Y < \tau_A]} \leq \frac{1}{1 - \mathbb{P}_{\nu_{X,A}}[\tau_A < \tau_Y]}. \quad (1.115)$$

The assertion of the lemma follows once we have proven that $\mathbb{P}_{\nu_{X,A \cup Y}}[\tau_A < \tau_Y] \leq \delta$ and $\mathbb{P}_{\nu_{Y,X}}[\tau_A < \tau_X] \leq \delta$. From the renewal equation (1.89) we immediately get that

$$\mathbb{P}_{\nu_{X,A \cup Y}}[\tau_A < \tau_Y] \leq \frac{\mathrm{cap}(X, A)}{\mathrm{cap}(X, Y)} \leq \delta. \quad (1.116)$$

On the other hand a further application of the seconds Green's identity to the functions $h_{A,X}$ and $h_{Y,X}$ shows that

$$\sum_{\sigma \in A} \mu_N(\sigma)\, (Lh_{Y,X})(\sigma)\, h_{A,X}(\sigma) = \sum_{\sigma \in Y} \mu_N(\sigma)\, (Lh_{A,X})(\sigma)\, h_{Y,X}(\sigma) \quad (1.117)$$

which is equivalent to

$$\mathbb{P}_{\nu_{Y,X}}[\tau_A < \tau_X] = \frac{\mathrm{cap}(X, A)}{\mathrm{cap}(X, Y)}\, \mathbb{P}_{\nu_{A,X}}[\tau_Y < \tau_X] \leq \delta. \quad (1.118)$$

This proves (1.113). □

1.9. Almost ultrametricity of capacities

Proof of Lemma 1.26 via probabilistic arguments. In order to prove a lower bound, consider the Markov process with initial distribution μ_A. Then

$$\mathbb{P}_{\mu_A}[\tau_Y < \tau_A] = \mathbb{P}_{\mu_A}[\tau_Y < \tau_A, \tau_X < \tau_A] + \mathbb{P}_{\mu_A}[\tau_Y < \tau_A, \tau_A < \tau_X]$$
$$\leq \mathbb{P}_{\mu_A}[\tau_X < \tau_A] + \sum_{\eta \in Y} \mathbb{P}_{\mu_A}[\tau_Y < \tau_{A \cup X}, \sigma(\tau_Y) = \eta] \, \mathbb{P}_\eta[\tau_A < \tau_X]. \tag{1.119}$$

By reversibility, we have for all $\eta \in Y$ that

$$\sum_{\sigma \in A} \mu_N(\sigma) \, \mathbb{P}_\sigma[\tau_Y < \tau_{A \cup X}, \sigma(\tau_Y) = \eta] = \mu_N(\eta) \, \mathbb{P}_\eta[\tau_A < \tau_{Y \cup X}]$$
$$\leq \mu_N(\eta) \, \mathbb{P}_\eta[\tau_A < \tau_Y]. \tag{1.120}$$

Hence, (1.119) can be rewritten as

$$\mathbb{P}_{\mu_A}[\tau_Y < \tau_A] \leq \mathbb{P}_{\mu_A}[\tau_X < \tau_A] + \frac{1}{\mu_N[A]} \sum_{\eta \in Y} \mu_N(\eta) \, \mathbb{P}_\eta[\tau_A < \tau_Y] \, \mathbb{P}_\eta[\tau_A < \tau_X]$$

which is equivalent to

$$\mathbb{P}_{\nu_{Y,A}}[\tau_X < \tau_A] = \sum_{\eta \in Y} \frac{\mu_N(\eta) \, \mathbb{P}_\eta[\tau_A < \tau_Y]}{\text{cap}(Y, A)} \mathbb{P}_\eta[\tau_X < \tau_A] \leq \frac{\text{cap}(X, A)}{\text{cap}(Y, A)}. \tag{1.121}$$

By a similar computation, we derive that $\mathbb{P}_{\nu_{Y,X}}[\tau_A < \tau_X] \leq \text{cap}(X, A) / \text{cap}(Y, X)$. Hence,

$$\frac{\text{cap}(X, A)}{\text{cap}(Y, A)} \geq 1 - \mathbb{P}_{\nu_{Y,A}}[\tau_A < \tau_X] \geq 1 - C_2 \frac{\text{cap}(X, A)}{\text{cap}(Y, X)} \geq 1 - C_2 \delta, \tag{1.122}$$

where we used that $\text{cap}(X, A) \leq \delta \, \text{cap}(X, Y)$ by assumption. To obtain an upper bound on the ratio of $\text{cap}(X, A)$ and $\text{cap}(Y, A)$, we repeat the computation leading to (1.121) with Y and X interchanged. This gives

$$\mathbb{P}_{\nu_{X,A}}[\tau_Y < \tau_A] \leq \frac{\text{cap}(Y, A)}{\text{cap}(X, A)}. \tag{1.123}$$

On the other hand, the averaged renewal equation (1.89) implies

$$\mathbb{P}_{\nu_{X, B \cup M}}[\tau_B < \tau_M] \leq \frac{\text{cap}(X, B)}{\text{cap}(M, B)} \leq \delta. \tag{1.124}$$

Hence, by combing (1.124) with (1.123) we finally get that

$$\frac{\text{cap}(X, A)}{\text{cap}(Y, A)} \leq \left(1 - \mathbb{P}_{\nu_{X,A}}[\tau_A < \tau_Y]\right)^{-1}$$
$$\leq \left(1 - C_1 \, \mathbb{P}_{\nu_{X, A \cup Y}}[\tau_A < \tau_Y]\right)^{-1}$$
$$\leq (1 - C_1 \delta)^{-1}. \tag{1.125}$$

This completes the proof. $\square$

Lemma 1.27. Let $X, A \subset \mathcal{S}_N$ be disjoint and $Y \subsetneq X$. Assume that there exists $0 < \delta < \frac{1}{C}$ with $C = \mathbb{P}_{\nu_{X,A}}[\tau_A < \tau_Y]/\mathbb{P}_{\nu_{X\setminus Y,Y}}[\tau_A < \tau_Y]$ such that $\mathrm{cap}(Y, A) \leq \delta \, \mathrm{cap}(Y, X\setminus Y)$. Then

$$1 - C\delta \leq \frac{\mathrm{cap}(Y, A)}{\mathrm{cap}(X, A)} \leq 1. \tag{1.126}$$

Proof. By applying the second Green's identity to the functions $h_{Y,A}$ and $h_{X,A}$ we get

$$\sum_{\sigma \in Y} \mu_N(\sigma) \, (Lh_{Y,A})(\sigma) = \sum_{\sigma \in X} \mu_N(\sigma) \, (Lh_{X,A})(\sigma) \, h_{Y,A}(\sigma).$$

Using the property that the harmonic function $h_{Y,A}$ is bounded from above by one as well as the definition of capacities, the upper bound in (1.126) is immediate. Moreover, by replacing $h_{Y,A}(\sigma)$ by $\mathbb{P}_\sigma[\tau_Y < \tau_A]$ for all $\sigma \in Y$, provides the following lower bound

$$\frac{\mathrm{cap}(Y, A)}{\mathrm{cap}(X, A)} \geq \mathbb{P}_{\nu_{X,A}}[\tau_Y < \tau_A] = 1 - C \, \mathbb{P}_{\nu_{X\setminus Y,Y}}[\tau_A < \tau_Y].$$

Hence, it remains to show that $\mathbb{P}_{\nu_{X\setminus Y,Y}}[\tau_A < \tau_Y] \leq \delta$. Now, a further application of the second Green's identity to the functions $h_{X\setminus Y,Y}$ and $h_{A,Y}$ shows that

$$\sum_{\sigma \in X\setminus Y} \mu_N(\sigma) \, h_{A,Y}(\sigma) \, (Lh_{X\setminus Y,Y})(\sigma) = \sum_{\sigma \in A} \mu_N(\sigma) \, h_{X\setminus Y,Y}(\sigma) \, (Lh_{A,Y})(\sigma)$$

which is equivalent to

$$\mathbb{P}_{\nu_{X\setminus Y,Y}}[\tau_A < \tau_Y] = \frac{\mathrm{cap}(Y, A)}{\mathrm{cap}(Y, X\setminus Y)} \, \mathbb{P}_{\nu_{A,Y}}[\tau_{X\setminus Y} < \tau_Y] \leq \delta.$$

This completes the proof of (1.126). $\square$

Corollary 1.28. Let $X, Y, A \subset \mathcal{S}_N$ be mutually disjoint, C_2 be given via (1.112) and consider $0 < \delta < \frac{1}{1+C_2}$. Then,

$$\mathrm{cap}(X, A) \geq \delta \min\{\mathrm{cap}(X, Y), \mathrm{cap}(Y, A)\}. \tag{1.127}$$

Proof. By contradiction. Assume that $\mathrm{cap}(X, A) < \delta \min\{\mathrm{cap}(X, Y), \mathrm{cap}(Y, A)\}$. This implies in particular that $\mathrm{cap}(X, A) < \delta \, \mathrm{cap}(X, Y)$. Hence, by Lemma 1.26 it holds that

$$\mathrm{cap}(Y, A) \leq \frac{1}{1 - C_2 \delta} \, \mathrm{cap}(X, A) \leq \frac{\delta}{1 - C_2 \delta} \, \mathrm{cap}(Y, A)$$

which is in contradiction with the assumption $\delta < \frac{1}{1+C_2}$. $\square$

Chapter 2

Coupling arguments and pointwise estimates

In the previous chapter we have presented the first important ingredient of the potential theoretic approach to metastability. Namely, a formula that relates the averaged mean hitting time of a set B when the process is started in a specific measure, $\nu_{A,B}$, on the set A with the capacity between these sets and the corresponding equilibrium potential. An natural question that comes to mind is whether the mean hitting time of B really depend on $\nu_{A,B}$.

This chapter has three main objectives. The first is to couple two versions of the Markov chain $\{\sigma(t)\}$, that starts in different configurations. For later applications the coupling should be constructed in such a way that allows to control the probability that both chains have be merged until a given time T. For this purpose, we first describe in Section 2.1 a general setting in which our methods can be applied. Notice that this setting is slightly more specific compared to the one considered in the previous chapter. Afterwards, we present in Section 2.2 the actual construction of the coupling. Let us point out that this tool is crucial in the later investigations. In Section 2.3 we address the second objective. We prove an estimate on the oscillations of harmonic functions within mesoscopic sets. Although the resulting bounds on equilibrium potentials are quiet rough, they are sufficient to compute precisely e.g. averaged metastable exit times in the next chapter. Let us point out that a rough estimate of the regularity of harmonic functions within mesoscopic sets is the second important ingredient of our approach. The third objective of this chapter is to derive sharp of mean hitting times and its Laplace transform. In situations where the Markov chain returns often to a small neighborhood of its starting point $\sigma \in A$ before escaping to the set B, one expects that mean exit times are almost constant as functions on the starting configuration. By taking advantage of this kind of recurrence property, in Section 2.4 we decompose the path of a Markov chain into cycles, and we prove as a first step a lower bound for the probability that the Markov chain returns to A before escaping to the set B. As presented in Section 2.4 this cycle decomposition is the tool in order to establish in a second step the desired sharp estimates.

2.1. Introduction

Coupling methods are well-known in all branches of probability theory and many textbooks on stochastic processes include this useful tool [**110**, **82**]. Since we use this method heavily in this chapter, it might be a good idea to recall its definition.

Definition 2.1 (Coupling). A coupling of two probability distributions μ and ν is a pair of random variables (X, Y) defined on a single probability space $(\Omega, \mathcal{F}, \mathbb{P})$ such that the marginal distribution of X is μ and the marginal distribution of Y is ν. That is, a coupling (X, Y) satisfies $\mathbb{P}\left[X = \sigma\right] = \mu(\sigma)$ and $\mathbb{P}\left[Y = \eta\right] = \nu(\eta)$.

In particular, we will consider a Markovian coupling of two Markov chains $\{\sigma(t)\}$ and $\{\eta(t)\}$ on $\mathcal{S}_N$, having the same transition probabilities $p_N(\sigma, \eta)$, which start in two different initial distributions. That is, a Markov chain $\{(\sigma(t), \eta(t))\}$ on $\mathcal{S}_N \times \mathcal{S}_N$ such that the marginal processes have transition probabilities $p_N(\sigma, \eta)$. Of course, such a coupling will only be useful if we introduce a coupling mechanism that connects both Markov chains in a non-trivial way.

Setting. We consider a Markov chain $\{\sigma(t)\}_{t \in \mathbb{N}_0}$ with transition probabilities, denoted by p_N, on a finite state space $\mathcal{S}_N = \{1, \ldots, q\}^N$. Here, N is a large parameter. We assume that the process is *irreducible* and *reversible* with respect to the unique stationary (Gibbs) measure μ_N. The transition probabilities p_N are such that the chain should evolve by selecting at each step a site $i \in \{1, \ldots, N\}$ uniformly at random and setting the corresponding spin variable to a randomly chosen $r \in \mathcal{S}_0 = \{1, \ldots, q\}$ according to the distribution $N p_N(\sigma, \sigma^{i,\cdot})$ on $\mathcal{S}_0$. Elements of $\mathcal{S}_0$ will be called *colors*. Moreover, we assume that there exists $\alpha > 0$ such that

$$N p_N(\sigma, \sigma^{i,r}) \geq \alpha \qquad \forall \sigma \in \mathcal{S}_N, \quad i \in \{1, \ldots, N\} \quad \text{and} \quad r \in \mathcal{S}_0. \tag{2.1}$$

Here, the configuration $\sigma^{i,r}$ is obtained from σ by replacing the color σ_i at site i through $r \in \mathcal{S}_0$. Its discrete generator, L_N, acts on functions $f \colon \mathcal{S}_N \to \mathbb{R}$ as

$$(L_N f)(\sigma) = \sum_{\eta \in \mathcal{S}_N} p_N(\sigma, \eta) \left(f(\eta) - f(\sigma)\right).$$

2.1. Introduction

Moreover, for a given subset $X \subset \mathcal{S}_N$ we introduce the notation $\mu_X(\sigma) = \mu_N[\sigma \,|\, X]$ for $\sigma \in X$ to denote the reversible measure, μ_N, conditioned on X.

Given a sequence of partitions $\{\Lambda^n\}_{n\in\mathbb{N}} \equiv \{\Lambda_1^n, \ldots, \Lambda_{k_n}^n\}_{n\in\mathbb{N}}$ of $\{1, \ldots, N\}$ having the property that Λ^{n+1} is a refinement of Λ^n, we consider a family of maps $\varrho^n : \mathcal{S}_N \to \Gamma^n \subset \mathbb{R}^{k_n \cdot q}$

$$\varrho^n(\sigma) := \sum_{k=1}^{k_n} e^k \otimes \frac{1}{N} \sum_{i\in\Lambda_k^n} \delta_{\sigma_i} \qquad (2.2)$$

where $e^k \in \mathbb{R}^{k_n}$ denotes the coordinate vector in $\mathbb{R}^{k_n}$ and δ_x is the point-mass at $x \in \mathbb{R}$. One may think of the maps ϱ^n as a vector of averages of *microscopic variables* σ_i over blocks of *mesoscopic sizes* which are decreasing in n.

Notice that the image process $\{\varrho^n(\sigma(t))\}$ on Γ^n is in general not Markovian. However, there is a canonical Markov process $\{\varrho^n(t)\}$ with state space Γ^n which is reversible with respect to the measure $\mathcal{Q}^n := \mu_N \circ (\varrho^n)^{-1}$ having the property that $\varrho^n(t) = \varrho^n(\sigma(t))$ in law whenever $\{\varrho^n(\sigma(t))\}$ is a Markov process. Its transition probabilities are given by

$$r^n(x, y) := \frac{1}{\mathcal{Q}^n(x)} \sum_{\sigma \in \mathcal{S}^n[x]} \mu_N(\sigma) \sum_{\eta \in \mathcal{S}^n[y]} p_N(\sigma, \eta), \qquad x, y \in \Gamma^n. \qquad (2.3)$$

To simplify notation, we denote by $\mathcal{S}^n[x] := (\varrho^n)^{-1}(x)$ the set-valued preimage of ϱ^n.

Now, the key assumption is to choose the sequence of partitions $\{\Lambda^n\}$ in such a way that the following two conditions holds true:

(A.1) There exists $\varepsilon(n) \downarrow 0$ as $n \uparrow \infty$ such that for any $x, y \in \Gamma^n$ with $r^n(x, y) > 0$,

$$\max_{\sigma \in \mathcal{S}^n[x], \eta \in \mathcal{S}^n[y]} \left| \frac{p_N(\sigma, \eta) \,|\{\eta \in \mathcal{S}^n[y] \mid d_{\mathrm{H}}(\sigma, \eta) = 1\}|}{r^n(x, y)} - 1 \right| \leq \frac{\varepsilon(n)}{3}, \qquad (2.4)$$

(A.2) If $\varrho^n(\sigma) = \varrho^n(\eta)$ and $\sigma_i = \eta_i$, for some i, then $p_N(\sigma, \sigma^{i,r}) = p_N(\eta, \eta^{i,r})$ for all $r \in \mathcal{S}$.

Let us remark that, provided both conditions above are satisfied, the Markov process $\{\varrho^n(t)\}$ can be seen as a good approximation of the process $\{\varrho^n(\sigma(t))\}$.

Our starting point in the study of the dynamical behavior of $\{\sigma(t)\}$ will be the following notion of *mimics a metastable situation on level n*.

Definition 2.2 (Metastable situation). Let us consider two disjoint subsets $A, B \subset \mathcal{S}_N$ for which there exists $n \in \mathbb{N}$ such that A and B are given as the preimage of some $\boldsymbol{A}, \boldsymbol{B} \subset \Gamma^n$ under ϱ^n. We say that A, B mimics a metastable situation of level n if there exists constants $0 < c < C$ such that

(i) for all $X = \mathcal{S}^n[\boldsymbol{x}]$ with $\boldsymbol{x} \in \boldsymbol{A}$

$$\mathbb{P}_{\mu_X}\left[\tau_B < \tau_X\right] \leq e^{-CN}, \tag{2.5}$$

(ii) for all $X = \mathcal{S}^n[\boldsymbol{x}], Y = \mathcal{S}^n[\boldsymbol{y}]$ with $\boldsymbol{x} \neq \boldsymbol{y} \in \boldsymbol{A}$

$$\mathbb{P}_{\mu_X}\left[\tau_Y < \tau_X\right] \geq e^{-cN}. \tag{2.6}$$

In the sequel, we will make the dependence on n explicit whenever we want to stress it. Otherwise, we will drop the superscript n, identify $k_n \equiv n$ and refer to the generic partition $\Lambda_1, \ldots, \Lambda_n$.

The challenge. In the context of stochastic spin systems, we are typically faced this the following situation. By (2.5), the Markov chain $\{\sigma(t)\}$ starting in $\sigma \in A$ will spent an exponential large amount of time in the neighborhood of A. On the other hand, the set A contains an exponentially large number of configurations and for any given configurations $\sigma, \eta \in A$, the probability of the Markov chain that starts in σ to hit η before entering in the set B is exponential small as $N \to \infty$. Hence, $\{\sigma(t)\}$ visits not more than a small fraction of A. In particular, the „time of thermalization", i.e. the typical time of two consecutive visits of a single configuration $\sigma \in A$ before entering in the set B, is of the same order than the metastable exit times. Thus, we can not expect that the chain equilibrates in a neighborhood of A before reaching B. In a recent paper [5], Bianch, Bovier and Ioffe have presented a strategy to derive pointwise estimates of mean hitting times in such a situation for a certain class of Ising-spin system, i.e. $\mathcal{S}_N = \{-1, 1\}^N$. Their approach is based on an explicit construction of a coupling of two Markov chains $\{\eta(t)\}$ and $\{\sigma(t)\}$ starting in different configurations $\sigma, \eta \in A$.

In what follows, we let us present the problem, we are faced with, in the construction of a coupling and sketch the main ideas and arguments to overcome these difficulties.

Let σ and η be two configurations with $\varrho^n(\sigma) = \varrho^n(\eta)$. We will construct a coupling $\{(\sigma(t), \eta(t))\}$ of two microscopic Markov chains starting from initial configuration σ and η, respectively, such that $\{\sigma(t)\}$ as well as $\{\eta(t)\}$ are versions of the original Markov chain with transition matrix P_N. For the application, we have in mind, the coupling should have the property that the Hamming distance between $\sigma(t)$ and $\eta(t)$ is non-increasing as long as both microscopic chains have the same mesoscopic value. However, due to the assumption (A.2), when σ, η and $i, j \in \Lambda_k$, for any k, are such that $\varrho^n(\sigma) = \varrho^n(\eta)$ and $\sigma_i = \eta_j$ then

$$p_N(\sigma, \sigma^{i,r}) = p_N(\eta, \eta^{j,r}) \quad \Longleftrightarrow \quad i = j. \tag{2.7}$$

Hence, we cannot find a coupling which assures that both chains maintain always the same mesoscopic value. Whenever, for the first time one chain accept the proposed single site update while the other one does not, we will use the independent coupling afterwards.

2.2. Construction of the Coupling

On the other hand, a closer look at (2.4) reveals that the difference between the probabilities is quite small, in the sense that there exists n large enough such that for all σ, η with $\varrho^n(\sigma) = \varrho^n(\eta)$ and $i, j \in \Lambda_k$, for any $k \in \{1, \ldots, n\}$, with $\sigma_i = \eta_j$, it holds

$$\frac{p_N(\eta, \eta^{j,r})}{p_N(\sigma, \sigma^{i,r})} \leq e^{\varepsilon(n)}, \qquad \forall r \in \mathcal{S}_0. \tag{2.8}$$

Clearly, our strategy should be to merge both microscopic chains as long as their mesoscopic values coincide. But notice that such a procedure may involve an implicit sampling of η-paths which may distort their statistical properties.

The main idea to overcome this difficulty is to separate path properties of the η-chain from the probability to decide whether $\sigma(t)$ maintain the same mesoscopic value as $\eta(t)$. Moreover, by decomposing the trajectory of $\{\sigma(t)\}$ into cycles, we can iterate the coupling using each time an independent copy of the η-chain.

2.2. Construction of the Coupling

Let σ and η be two configurations with $\varrho^n(\sigma) = \varrho^n(\eta)$. Our first objective in this section is to explain how to couple the probability distributions $N p_N(\eta, \eta^{i,\cdot})$ and $N p_N(\sigma, \sigma^{j,\cdot})$ on $\mathcal{S}_0$ where i, j are chosen from some Λ_k such that $\sigma_i = \eta_j$. Taking advantage of (2.8), the coupling (X, Y) is constructed in such a way that we can decide in advance by tossing a coin whether Y realizes at least the same value as X.

Note that the actual construction of the coupling is a modification of the *optimal coupling* between two probability measures on a finite state space, as given in [82], which was already used in the easier case $|\mathcal{S}_0| = 2$ in [5].

Lemma 2.3. *Let $\mu, \nu \in \mathcal{M}_1(\mathcal{S}_0)$. Suppose that there exits $\delta \in (0, 1)$ such that $\delta \mu(r) \leq \nu(r)$ for all $r \in \mathcal{S}_0$. Then there exists an optimal coupling (X, Y) of μ and ν with the additional property that there exists a Bernoulli-δ-distributed random variable U independent of X such that*

$$\mathbb{P}[Y = s \mid U = 1, X = r] = \mathbb{1}_{r=s}. \tag{2.9}$$

Proof. To start with, choose a color $r \in \mathcal{S}_0$ randomly according to μ and set $X = r$. We use the following procedure to generate Y. Since $X = r$ implies that $\mu(r) > 0$

$$p(r) = \frac{\mu(r) \wedge \nu(r) - \delta \mu(r)}{(1 - \delta) \mu(r)} \tag{2.10}$$

is well defined. Now, flip a coin with probability of heads equal to δ. If the coin comes up heads, then set $Y = r$. Otherwise, flip another coin with probability of heads equal to $p(r)$. If the second coin comes up heads, set again $Y = r$. If it comes up tails, choose

Y according to the probability distribution γ_r on $\mathcal{S}_0 \setminus \{r\}$ given by

$$\gamma_r(s) = \frac{(\nu(s) - \mu(s))\, \mathbb{1}_{\nu(s) \geq \mu(s)}}{\sum_{z \in \mathcal{S}_0 \setminus \{r\}} (\nu(z) - \mu(z))\, \mathbb{1}_{\nu(z) \geq \mu(z)}}. \tag{2.11}$$

By construction, the additional property in the statement of the lemma is immediate. It remains to check that Y is indeed distributed according to ν. Hence, consider

$$\mathbb{P}[Y = r] = \delta\,\mu(r) + (1 - \delta)\,\mu(r)\,p(r) + \sum_{s \in \mathcal{S}_0 \setminus \{r\}} (1 - \delta)\,\mu(s)\,(1 - p(s))\,\gamma_s(r)$$

$$= \mu(r) \wedge \nu(r)$$

$$+ (\nu(r) - \mu(r) \wedge \nu(r)) \sum_{s \in \mathcal{S}_0 \setminus \{r\}} \frac{(\mu(s) - \nu(s))\,\mathbb{1}_{\mu(s) \geq \nu(s)}}{\sum_{z \in \mathcal{S}_0 \setminus \{s\}} (\nu(z) - \mu(z))\,\mathbb{1}_{\nu(z) \geq \mu(z)}} \tag{2.12}$$

Expressing the denominator in (2.12) in terms of the *total variation distance* reveals that

$$\sum_{\substack{z \in \mathcal{S}_0 \setminus \{s\} \\ \nu(z) \geq \mu(z)}} (\nu(z) - \mu(z)) = \|\mu - \nu\|_{\mathrm{TV}} - (\nu(s) - \mu(s))\,\mathbb{1}_{\nu(s) \geq \mu(s)}. \tag{2.13}$$

Thus, by distinguishing two cases, this implies that the sum in (2.12) equals

$$\sum_{s \in \mathcal{S}_0 \setminus \{r\}} \frac{(\mu(s) - \nu(s))\,\mathbb{1}_{\mu(s) \geq \nu(s)}}{\|\mu - \nu\|_{\mathrm{TV}}} = 1 - \frac{(\mu(r) - \nu(r))\,\mathbb{1}_{\mu(r) \geq \nu(r)}}{\|\mu - \nu\|_{\mathrm{TV}}} \tag{2.14}$$

But notice that, since $\nu(r) - \mu(r) \wedge \nu(r) = (\nu(r) - \mu(r))\,\mathbb{1}_{\nu(r) \geq \mu(r)}$, the last term in (2.14) vanishes. Hence the probability in (2.12) equals $\nu(r)$, as desired. In order to prove the optimality of the coupling, it remains to show that $\mathbb{P}[X \neq Y] = \|\mu - \nu\|_{\mathrm{TV}}$. Note that the coupling procedure implies that

$$\mathbb{P}[X \neq Y] = 1 - \sum_{r \in \mathcal{S}_0} \bigl(\delta\,\mu(r) + (1 - \delta)\,\mu(r)\,p(r)\bigr)$$

$$= 1 - \sum_{r \in \mathcal{S}_0} \mu(r) \wedge \nu(r)$$

$$= \|\mu - \nu\|_{\mathrm{TV}}, \tag{2.15}$$

where we used in the last step that

$$\sum_{r \in \mathcal{S}_0} \mu(r) \wedge \nu(r) = \sum_{\substack{r \in \mathcal{S}_0 \\ \mu(r) < \nu(r)}} \mu(r) + \sum_{\substack{r \in \mathcal{S}_0 \\ \mu(r) \geq \nu(r)}} \nu(r) = 1 - \|\mu - \nu\|_{\mathrm{TV}}. \tag{2.16}$$

This completes the proof of the lemma. □

2.2. Construction of the Coupling

Let us now explain one step of a coupling that was used by [5] in the study of the random field Curie-Weiss model $q = 2$. Note that the coupling mechanism was originally invented in [81]. Recall that we consider σ, η with $\varrho^n(\sigma) = \varrho^n(\eta)$.

To update both configurations pick a site $i \in \{1, \ldots, N\}$ uniformly at random. Whenever η_i and σ_i coincides, (2.7) guarantees that there is a coupling assuring that the σ-chain maintain the same mesoscopic values as the η-chain. On the other hand, if for some k and $i \in \Lambda_k$ the spin variables η_i and σ_i differs, the idea is to choose j uniformly at random among all sites in Λ_k having the property that $\eta_j \neq \sigma_j$ and $\eta_i = \sigma_j$ and to apply Lemma 2.3 to the probability distributions $N p_N(\eta, \eta^{i,\cdot})$ and $N p_N(\sigma, \sigma^{j,\cdot})$. By tossing a coin with success probability $e^{-\varepsilon(n)}$, the coupling, used in Lemma 2.3, allows to decide in advance and independently on what happens to the η-chain whether both chains at least maintain the property having the same mesoscopic value.

Let us emphasis the fact that, provided the coin comes up heads, the Hamming distance between η and σ decreases by one whenever a site i is updated where $\sigma_i \neq \eta_i$. This allows to define an event, depending only on the filtration induced by the η-chain and the outcome of the coin tosses, which guaranties the coalescence of both chains. However, this property is no longer true in the general case $q \geq 3$, although the coupling, described above, could also be applied there.

In order to overcome this problem, we modify the coupling above by taking into account a family of bijections π_t between the sites in the configurations $\eta(t)$ and $\sigma(t)$, respectively. Namely, for the initial configurations η and σ let π_0 be a permutation of $\{1, \ldots, N\}$ with the property that

(i) $\pi_0(i) = i$, if and only if, $\sigma_i = \eta_i$,
(ii) $\pi_0(i) = j$, if and only if, there exists $k \in \{1, \ldots, n\}$ such that $i, j \in \Lambda_k$ and

$$\eta_i \neq \sigma_i \quad \wedge \quad \eta_j \neq \sigma_j \quad \wedge \quad \eta_i = \sigma_j.$$

Since we have assumed that $\varrho^n(\sigma) = \varrho^n(\eta)$, π_0 is well defined. Moreover, for any $t > 1$, the construction of π_t depends only on the filtration induced by the Markov chain started in η. In order to see this, suppose π_t is already given and at time $t+1$ site i is drawn. If $\pi_t(i) = i$ set $\pi_{t+1} = \pi_t$. Otherwise, we have to distinguish the following four cases. If

(1) $\eta_i(t+1) \neq \eta_{\pi_t(i)}(t)$ and $\eta_i(t+1) \neq \eta_{\pi_t^{-1}(i)}(t)$ set $\pi_{t+1} = \pi_t$.
(2) $\eta_i(t+1) = \eta_{\pi_t(i)}(t)$ and $\eta_i(t+1) \neq \eta_{\pi_t^{-1}(i)}(t)$ set

$$\pi_{t+1}(\pi_t(i)) = \pi_t(i), \quad \pi_{t+1}(i) = \pi_t(\pi_t(i)), \quad \pi_{t+1}(j) = \pi_t(j), \; \forall j \neq i, \pi_t(i).$$

(3) $\eta_i(t+1) \neq \eta_{\pi_t(i)}(t)$ and $\eta_i(t+1) = \eta_{\pi_t^{-1}(i)}(t)$ set

$$\pi_{t+1}(i) = i, \quad \pi_{t+1}(\pi_t^{-1}(i)) = \pi_t(i), \quad \pi_{t+1}(j) = \pi_t(j), \; \forall j \neq i, \pi_t^{-1}(i).$$

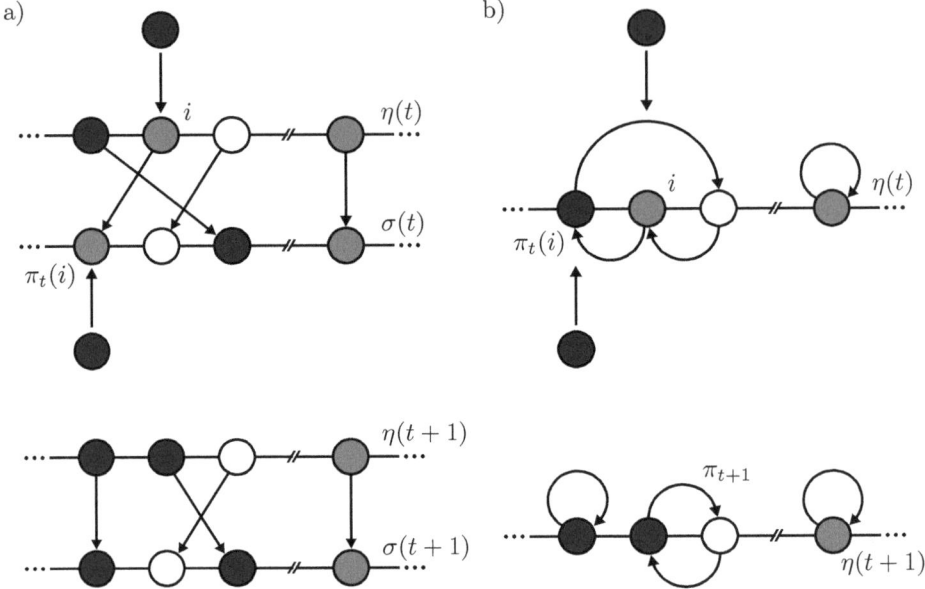

Figure 1. Illustration of an update step of $\eta(t)$ and $\sigma(t)$. a) Possible choice of a bijection between sites in the η- and σ-chain and its update in cases (2). b) Corresponding graphical representation in terms of $\eta(t)$ and π_t. On the event that all coin tosses involved in the construction come up heads, a merging of the η- and σ-chain is indicated by the fact that π_t is the identity, i.e. π_t is represented only by self-loops.

(4) $\eta_i(t+1) = \eta_{\pi_t(i)}(t)$ and $\eta_i(t+1) = \eta_{\pi_t^{-1}(i)}(t)$ set

$$\pi_{t+1}(i) = i, \quad \pi_{t+1}(\pi_t(i)) = \pi_t(i), \quad \pi_{t+1}(\pi_t^{-1}(i)) = \pi_t(\pi_t(i)),$$

and $\pi_{t+1}(j) = \pi_t(j)$ for all $j \neq i, \pi_t(i), \pi_t^{-1}(i)$.

An illustration of construction of π_{t+1} from π_t is given in Figure 2.2.

Now, we describe a coupling of two Markov processes starting from the initial configuration η and σ. Recall that η and σ satisfy $\varrho^n(\eta) = \varrho^n(\sigma)$. There are two parameters M and T involved those values will be quantified later on. Let $\{V_0, \ldots, V_{M-1}\}$ be a sequence of i.i.d. Bernoulli random variables with

$$\mathbb{P}[V_i = 1] = 1 - \mathbb{P}[V_i = 0] = e^{-\varepsilon(n)}. \tag{2.17}$$

Now, set $\eta(0) = \eta$, $\sigma(0) = \sigma$, $M_0 = 0$ and $\xi = 0$. The coupling during the first T-steps of its life is given by

2.2. Construction of the Coupling

for $t = 0, 1, \ldots, T-1$ **do**
    if $\xi = 0 \wedge M_t < M$ **then**
        Choose $i \in \{1, \ldots, N\}$ uniform at random.
        if $\pi_t(i) = i$ **then**
            Choose $s \in S_0$ at random according to the distribution $N\, p_N(\eta, \eta^{i,s})$.
            Set

$$\sigma_j(t+1) = \begin{cases} \sigma_j(t), & j \neq i \\ s, & j = i \end{cases} \quad \text{and} \quad \eta_j(t+1) = \begin{cases} \eta_j(t), & j \neq i \\ s, & j = i \end{cases}.$$

            Set $M_{t+1} = M_t$ and $\pi_{t+1} = \pi_t$.
        else
            Apply the optimal coupling, describe in Lemma 2.3, to $N\, p_N(\eta(t), \eta(t)^{i,\cdot})$ and $N\, p_N(\sigma(t), \sigma(t)^{\pi_t(i),\cdot})$, where V_{M_t} decides if both chains maintain at least the same mesoscopic value.
            Set $M_{t+1} = M_t + 1$.
            if $V_{M_t} = 1$ **then**
                Generate π_{t+1} as described above.
            else
                Set $\xi = 1$.
            end if
        end if
    else
        Use the independent coupling to update $\eta(t)$ and $\sigma(t)$.
    end if
end for

For later reference, we denote by $\{I_t\}_{t=0}^{T-1}$ the family of independent and uniformly on $\{1, \ldots, N\}$ distributed random variables describing the random experiment of choosing a site in the algorithm. Moreover, notice that $\{M_t\}$ is a process that increases by one each time a new coin V_i is used in the coupling whereas the value $\xi = 1$ indicates that a coin comes up tails.

Lemma 2.4. *Let $\mathbb{P}_{\sigma,\eta}$ be the joint distribution of the processes $\{\eta(t)\}$, $\{\sigma(t)\}$ and $\{V_i\}$. Then, for all $t \leq T$, the construction above is a coupling of two Markov chains starting in η and σ with transition matrix P_N.*

Proof. The assertion is obvious as soon as $\xi = 1$ or $M_t \geq M$ for some $t < T$, since both chains are updated independently. Hence, it remains to check that the statement holds true when $\xi = 0$ and $M_t < M$. Due to the bijection, each lattice site is drawn in both chains with probability $1/N$. Moreover, in view of (2.7) and Lemma 2.3, we can conclude the statement of the lemma. □

In this construction, the Hamming distance between $\eta(t)$ and $\sigma(t)$ is non-increasing as long as $\xi = 0$ and $M_t < M$. Provided that for some $t < T$ these conditions still hold and both chains coalesce, $\eta(t) = \sigma(t)$, then the two dynamics stay automatically together until T. As it was pointed out in [5], conditioned on such a situation involves an implicit sampling of η-paths which may lead to a distortion of their statistical properties.

In order to avoid an implicit sampling of paths, we will define two *independent* events which allows to conclude that a merging of the Markov chains started in η and σ, respectively, has taken place whenever both events happen simultaneously. This strategy was successfully demonstrated in [5]. Hence, let us consider

(i) the event that all coin tosses has come up heads, i.e.

$$\mathcal{A} := \{V_i = 1, \quad \forall i \in \{0, \ldots, M-1\}\}. \tag{2.18}$$

(ii) an event depending only on the random variables $\eta(t)$ for $t < T$. In order to define it, notice that for any realization of $\{\eta(t)\}$ and given π_0, which is determined by the initial configuration σ, we can uniquely construct $\{\pi_t\}$. Moreover, we introduce stopping times, t_i, through

$$t_i := \inf\{t \geq 0 \mid \eta_i(t+1) \neq \eta_i(t) \wedge \eta_i(t+1) = \eta_{\pi_t^{-1}(i)}(t) \text{ if } \pi_t^{-1} \neq i\}$$

and set $\mathfrak{t} := \max_{1 \leq i \leq N} t_i$. By construction, t_i is the first time that the spin at site i has been flipped to a specific or an arbitrary color depending on $\{\pi_t\}$. Further define the random variable

$$\mathcal{N} := \sum_{i=1}^{N} \sum_{t=0}^{t_i} \mathbb{1}_{I_t = i}. \tag{2.19}$$

representing the total number of flipping attempts until time $\mathfrak{t}$. Finally, let us consider the events

$$\mathcal{B} := \{\tau_B \geq T\} \cap \{\mathfrak{t} < T\} \cap \{\mathcal{N} \leq M\}. \tag{2.20}$$

An important observation is the following

Lemma 2.5. *For any given value of the parameters T and M, on $\mathcal{A} \cap \mathcal{B}$ a merging of $\{\eta(t)\}$ and $\{\sigma(t)\}$ has taken place, in the sense that*

$$\mathcal{A} \cap \mathcal{B} \subset \{\eta(T) = \sigma(T)\}. \tag{2.21}$$

2.2. Construction of the Coupling

Proof. On the event $\mathcal{A} \cap \mathcal{B}$, the process $\{\eta(t)\}$ has not reached B by time T but all spins have been flipped at least once to an arbitrary or to a specific color depending on $\{\pi_t\}$. Notice that whenever a site i where $\eta_i \neq \sigma_i$ is flipped to the color specified by $\{\pi_t\}$, the corresponding spins become aligned because on $\mathcal{A} \cap \mathcal{B}$ all coin tosses come up heads. Since $t < T$ this implies that $\eta(T) = \sigma(T)$. □

Further, by taking advantage of the property (2.1), we can establish a bound on the tail probability of the random variable $\mathcal{N}$.

Lemma 2.6. *Let $c_1 > 1 + \alpha^{-1}$ and set $M = c_1 N$. Then, it holds that*

$$\mathbb{P}_\eta \left[\mathcal{N} > M \right] \leq e^{-c_2 N}. \tag{2.22}$$

with a constant $c_2 = -c_1 \ln\left(1 - \alpha^2\right) - \ln\left(1 + \alpha\right)$.

Proof. In view of (2.1), the idea behind the proof is to construct on some probability space $(\Omega, \mathcal{F}, \mathbb{P})$ a coupling between the Markov chain $\{\eta(t)\}$ and a sequence $\{\omega(t)\}$ of i.i.d. Bernoulli random variables in such a way that we can define a random variable $\mathfrak{N}$ depending only on the natural filtration induced by $\{\omega(t)\}$ with the property that $\mathcal{N} \leq \mathfrak{N}$. Clearly, by bounding the tail probability of $\mathfrak{N}$ from above, we obtain an upper bound on the tail probability $\mathcal{N}$.

Let us now describe one step of the actual coupling construction. At time $t+1$ ($t \geq 0$) choose a site $i \in \{1, \ldots, N\}$ uniformly at random and toss a coin with probability of heads equal to α. Depending on π_t, we have to distinguish two cases:

(1) $\pi_t(i) \neq i$. If the coin comes up heads, set $\omega(t+1) = 1$ and $\eta_i(t+1)$ to the color given by $\eta_{\pi_t^{-1}(i)}(t)$. Otherwise, set $\omega(t+1) = 0$ and draw a $r \in S_0$ according to the distribution

$$\gamma_{i,t}(r) = \begin{cases} \dfrac{N\, p_N\left(\eta(t), \eta(t)^{i,r}\right) - \alpha}{1 - \alpha}, & r = \eta_{\pi_t^{-1}(i)}(t) \\[2mm] \dfrac{N\, p_N\left(\eta(t), \eta(t)^{i,r}\right)}{1 - \alpha}, & r \neq \eta_{\pi_t^{-1}(i)}(t), \end{cases} \tag{2.23}$$

(2) $\pi_t(i) = i$. If the coin comes up heads, set $\omega(t+1) = 1$ and choose a color $r \in S_0 \setminus \{\eta_i(t)\}$ uniformly at random. Otherwise, set $\omega(t+1) = 0$ and draw a $r \in S_0$ according to the distribution

$$\gamma_{i,t}(r) = \begin{cases} \dfrac{N\, p_N\left(\eta(t), \eta(t)^{i,r}\right)}{1 - \alpha}, & r = \eta_i(t) \\[2mm] \dfrac{N\, p_N\left(\eta(t), \eta(t)^{i,r}\right) - \frac{\alpha}{q-1}}{1 - \alpha}, & r \neq \eta_i(t). \end{cases} \tag{2.24}$$

It is easy to check that the construction above is a coupling $\{(\eta(t),\omega(t))\}$ of the Markov chain $\{\eta(t)\}_{t\geq 1}$ and a sequence $\{\omega(t)\}_{t\geq 1}$ of independent Bernoulli-α-distributed random variables. Further, let us introduce the stopping time

$$\mathfrak{N} := \inf\left\{s \geq 1 \,\Big|\, \sum_{t=1}^{s} \omega(t) = N\right\} - N. \qquad (2.25)$$

Notice that $\mathfrak{N}$ is negative binomial distributed to the parameters N and α and, due to the construction above, it holds that $\mathcal{N} \leq \mathfrak{N} + N$. Hence, by standard large deviation estimates, we obtain

$$\mathbb{P}_{\eta}[\mathcal{N} > c_1 N] \leq \mathbb{P}[\mathfrak{N} > (c_1 - 1)N] \leq e^{-NI(c_1-1)} \leq e^{-c_2 N}. \qquad (2.26)$$

The last inequality uses the fact that the entropy I is convex and for all $c > \alpha^{-1}$

$$\begin{aligned} I(c) &= c\ln\frac{c}{1+c} - c\ln(1-\alpha) - \ln(1+c) - \ln\alpha \\ &\geq -\ln(1-\alpha^2)c - \ln(1+\alpha) \\ &> 0. \end{aligned}$$

This completes the proof. $\square$

2.3. Bounds on harmonic functions and local recurrence

As a first application of the coupling construction, we will show a statement about the regularity of harmonic functions within a given set of configurations that is determined by the same mesoscopic value.

Proposition 2.7. *For any $n \in \mathbb{N}$, let us consider two disjoint subsets $X, Y \subset \Gamma^n$ and denote by $\mathcal{X}, \mathcal{Y}$ their preimages under ϱ^n. Further, let $x \in \Gamma^n$ choose $c_1 > 1 + \alpha^{-1}$. Then,*

$$\mathbb{P}_{\sigma}[\tau_{\mathcal{X}} < \tau_{\mathcal{Y}}] \geq e^{-\varepsilon(n)c_1 N}\left(\mathbb{P}_{\eta}[\tau_{\mathcal{X}} < \tau_{\mathcal{Y}}] - e^{-c_2 N}\right), \qquad \forall\, \sigma, \eta \in \mathcal{S}^n[x], \qquad (2.27)$$

where $c_2 \equiv c_2(c_1)$ depends linearly on c_1.

Proof. We will use the coupling construction where we choose the involved parameters $T = \infty$ and $M = c_1 N$. Further, consider the following two events

$$\mathcal{B}_1 := \{\tau_{\mathcal{X}} < t\} \cap \{\mathcal{N} \leq M\} \quad \text{and} \quad \mathcal{B}_2 := \{\tau_{\mathcal{X}} \geq t\} \cap \{\mathcal{N} \leq M\}. \qquad (2.28)$$

Analog to Lemma 2.5, on the event $\mathcal{A} \cap \mathcal{B}_2$, the process $\{\eta(t)\}$ has not reached $\mathcal{X}$ by time t and at time t the spins at any site i with $\eta_i \neq \sigma_i$ become aligned. Since the corresponding coin tosses come up heads, this implies that $\eta(t) = \sigma(t)$ and hence $\tau_{\mathcal{X}}^{\eta} = \tau_{\mathcal{X}}^{\sigma}$. Moreover, on the event $\{\tau_{\mathcal{X}}^{\eta} < \tau_{\mathcal{Y}}^{\eta}\}$, it holds that $\tau_{\mathcal{Y}}^{\eta} = \tau_{\mathcal{Y}}^{\sigma}$, since X, Y are given as preimages under ϱ^n.

On the event $\mathcal{A} \cap \mathcal{B}_1$, the process $\{\eta(t)\}$ reaches $\mathcal{X}$ before time t. Since until time t all necessary coin tosses come up heads, we can conclude that $\varrho^n(\eta(t)) = \varrho^n(\sigma(t))$

2.4. Cycle decomposition of σ-paths

for all $t \leq \tau_X$. Therefore, due to the assumption on X, Y, it holds that $\tau_X^\eta = \tau_X^\sigma$ and on the event $\{\tau_X^\sigma < \tau_Y^\sigma\}$ the η-chain can not reach Y before time τ_X^η. Thus,

$$\begin{aligned}\mathbb{P}_\sigma\left[\tau_X < \tau_Y\right] &\geq \mathbb{P}_{\sigma,\eta}\left[\tau_X^\sigma < \tau_Y^\sigma, \mathcal{A} \cap \mathcal{B}_1\right] + \mathbb{P}_{\sigma,\eta}\left[\tau_X^\sigma < \tau_Y^\sigma, \mathcal{A} \cap \mathcal{B}_2\right]\\ &= \mathbb{P}_{\sigma,\eta}\left[\tau_X^\eta < \tau_Y^\eta, \mathcal{A} \cap \mathcal{B}_1\right] + \mathbb{P}_{\sigma,\eta}\left[\tau_X^\eta < \tau_Y^\eta, \mathcal{A} \cap \mathcal{B}_2\right]\\ &\geq \mathrm{e}^{-\varepsilon(n)M}\left(\mathbb{P}_\eta\left[\tau_X < \tau_Y\right] - \mathbb{P}_\eta\left[\mathcal{N} > M\right]\right).\end{aligned} \qquad (2.29)$$

Taking into account (2.22) concludes the proof of the lemma. □

Remark 2.8. Instead of (2.29), one can also consider the following lower bound which is based on the h-transform

$$\mathbb{P}_\sigma\left[\tau_X < \tau_Y\right] \geq \mathrm{e}^{-\varepsilon(n)M}\,\mathbb{P}_\eta\left[\tau_X < \tau_Y\right]\left(1 - \mathbb{P}_\eta^h\left[\mathcal{N} > M\right]\right).$$

In order to apply Lemma 2.6 to bound $\mathbb{P}_\eta^h[\mathcal{N} > M]$, it remains to show that $N\,p_N^h(\sigma, \eta)$ is bounded from below by a constant which is independent of N for all σ, η with $d_H(\sigma, \eta) \leq 1$. However, the only bound we can achieve is

$$N\,p_N^h(\sigma, \eta) = \frac{1}{\mathbb{P}_\sigma\left[\tau_X < \tau_Y\right]}\,N\,p_N(\sigma, \eta)\,\mathbb{P}_\eta\left[\tau_X < \tau_Y\right] \geq \frac{\mathbb{P}_\eta\left[\tau_X < \tau_Y\right]}{\mathbb{P}_\sigma\left[\tau_X < \tau_Y\right]}\,\alpha \geq \frac{\alpha^2}{N},$$

where we used in the last step that $\mathbb{P}_\eta\left[\tau_X < \tau_Y\right] \geq p_N(\eta, \sigma)\,\mathbb{P}_\sigma\left[\tau_X < \tau_Y\right]$.

Corollary 2.9. *Suppose that A, B mimics a metastable situation on level n. Consider $c_1 > 1 + \alpha^{-1}$ and assume that n satisfy $\varepsilon(n)c_1 < C$. Then, there exists $c_3 > 0$ such that, for N large enough,*

$$\mathbb{P}_\eta\left[\tau_B < \tau_A\right] \leq \mathrm{e}^{-c_3 N}, \qquad \forall \eta \in \mathcal{A}. \qquad (2.30)$$

Proof. Recall that the disjoint subsets $A, B \subset \mathcal{S}_N$ are given as preimages of $\mathcal{A}, \mathcal{B} \subset \Gamma^n$ under ϱ^n. Further, for any $\eta \in \mathcal{A}$ there exists $x \in A$ such that $\eta \in X = \mathcal{S}^n[x]$. Hence, by combining (2.5) and (2.27), we obtain

$$\begin{aligned}\mathbb{P}_\eta\left[\tau_B < \tau_X\right] &\leq \mathrm{e}^{\varepsilon(n)c_1 N}\,\mathbb{P}_{\mu_X}\left[\tau_B < \tau_X\right] + \mathrm{e}^{-c_2 N}\\ &\leq \mathrm{e}^{-N(C-\varepsilon(n)c_1)} + \mathrm{e}^{-c_2 N}\\ &\leq \mathrm{e}^{-c_3 N},\end{aligned}$$

with a constant $c_3 > 0$. Since $\mathbb{P}_\eta\left[\tau_B < \tau_A\right] \leq \mathbb{P}_\eta\left[\tau_B < \tau_X\right]$, we conclude the statement. □

2.4. Cycle decomposition of σ-paths

A further reason for introducing the likely event $\mathcal{B}$ is that it does not distort the hitting time of the Markov chain $\{\eta(t)\}$. This is the statement of the following lemma.

Lemma 2.10. *Uniformly for all $\eta \in A$, there exists a constant $c_4 > 0$, independent of n, such that for N large enough*

$$\mathbb{P}_\eta[\mathcal{B}^c] \leq e^{-c_4 N} \tag{2.31}$$

and

$$\mathbb{E}_\eta[\tau_B \mathbf{1}_\mathcal{B}] \geq \mathbb{E}_\eta[\tau_B]\left(1 - e^{-c_4 N}\right). \tag{2.32}$$

Proof. In order to prove this lemma, we follow in parts the proof of [5, Lemma 3.3]. We will use the coupling constructed above where we choose $T = N^\kappa$, for some $\kappa > 2$. To start with, notice that the event $\mathcal{B}^c$ can be written as

$$\begin{aligned}\mathcal{B}^c = {} & \{\tau_B \leq N^\kappa\} \cup \{\tau_B > N^\kappa\} \cap \{t \geq N^\kappa\} \\ & \cup \{\tau_B > N^\kappa\} \cap \{t < N^\kappa\} \cap \{\mathcal{N} > M\}.\end{aligned} \tag{2.33}$$

In view of (2.33), there are three terms to bound in the equations (2.31) and (2.32). Bounding the first term of (2.32) is quite simple. Namely,

$$\mathbb{E}_\eta\left[\tau_B \mathbf{1}_{\tau_B < N^\kappa}\right] \leq N^\kappa \, \mathbb{P}_\eta[\tau_B < N^\kappa]. \tag{2.34}$$

In order to reach B, we exploit the fact that the Markov chain $\{\eta(t)\}$ has to make one final excursion without returning to the starting set A and there are at most N^κ attempts. This implies that

$$\mathbb{P}_\eta[\tau_B \leq N^\kappa] \leq N^\kappa \max_{\sigma \in A} \mathbb{P}_\sigma[\tau_B < \tau_A] \leq N^\kappa e^{-c_3 N} \tag{2.35}$$

where we used in the last step the local recurrence property (2.30). The second term of (2.31) can be easily bounded by

$$\mathbb{P}_\eta[\tau_B > N^\kappa, t \geq N^\kappa] \leq \mathbb{P}_\eta[t \geq N^\kappa]. \tag{2.36}$$

By using the strong Markov property we can split the paths at time N^κ. By proceeding in such a way we obtain for the second term of (2.32)

$$\mathbb{E}_\eta\left[\tau_B \mathbf{1}_{\tau_B > N^\kappa} \mathbf{1}_{t \geq N^\kappa}\right] \leq \left(N^\kappa + \max_{\sigma \in S_N} \mathbb{E}_\sigma[\tau_B]\right) \mathbb{P}_\eta[t \geq N^\kappa] \leq e^{2c_5 N} \, \mathbb{P}_\eta[t \geq N^\kappa]. \tag{2.37}$$

Notice that, in order to obtain a rough upper bound on the mean hitting time $\mathbb{E}_\sigma[\tau_B]$, it is enough to bound the corresponding capacity $\mathrm{cap}(\sigma, B)$ from below. Following [15, 11], there exists $c_5 > 0$, independent of n, such that $\mathrm{cap}(\sigma, B) \geq e^{-c_5 N}$. Thus, it remains to show that the probability $\mathbb{P}_\eta[t \geq N^\kappa]$ is super-exponentially small. However, at each step the probability to flip a particular spin is bounded from below by α/N, which implies that

$$\mathbb{P}_\eta[t \geq N^\kappa] \leq N\left(1 - \frac{\alpha}{N}\right)^{N^\kappa} \leq e^{-c_6 N^{\kappa-1}}. \tag{2.38}$$

2.4. Cycle decomposition of σ-paths

Finally, we can bound the third term by

$$\mathbb{P}_\eta[\tau_B > N^\kappa, \mathfrak{t} < N^\kappa, \mathcal{N} > M] \leq \mathbb{P}_\eta[\mathcal{N} > M] \tag{2.39}$$

and, proceeding similarly as in the treatment of the second term, we obtain

$$\mathbb{E}_\eta[\tau_B \mathbb{1}_{\tau_B > N^\kappa} \mathbb{1}_{\mathfrak{t} < N^\kappa} \mathbb{1}_{\mathcal{N} > M}] \leq \left(N^\kappa + \max_{\sigma \in S_N} \mathbb{E}_\sigma[\tau_B]\right) \mathbb{P}_\eta[\mathcal{N} > M]$$

$$\leq e^{2c_5 N} \, \mathbb{P}_\eta[\mathcal{N} > M]. \tag{2.40}$$

For $M = c_1 N$ and $c_1 > 1+\alpha^{-1}$ Lemma 2.6 implies that $\mathbb{P}_\eta[\mathcal{N} > c_1 N] \leq e^{-c_2 N}$. Since c_2 increases linearly with c_1, the right-hand side of (2.40) is exponentially small, if we choose c_1 such that $c_2 > 2\,c_5$. Thus, by combining all estimates above we conclude the proof of the lemma. □

By Lemma 2.5, we know that the Markov chains $\{\sigma(t)\}$ and $\{\eta(t)\}$ has merged on the event $\mathcal{A} \cap \mathcal{B}$ by time T. While the probability of the event $\mathcal{B}$ is close to one, the probability of the event $\mathcal{A}$ is small, namely $e^{-\varepsilon(n)M}$. Hence, the occurrence of $(\mathcal{A} \cap \mathcal{B})^c$ is rather likely. However, in view of (2.5) and (2.30), respectively, the σ-chain will return to the set A after time T with a probability close to one. This allows to decompose $\{\sigma(t) : t \in [0, \tau_B^\sigma]\}$ into cycles.

Let us emphasis that the cycles decomposition, given below, goes along the lines of the construction originally presented in [5]. To start with, we define for any $\eta \in A$ the following stopping times

$$\mathfrak{s}_{-1} = 0, \qquad \mathfrak{s}_k := \inf\{t > \mathfrak{s}_{k-1} + T \mid \sigma(t) \in \varrho^n(\eta)\}, \qquad \forall k \in \mathbb{N}_0, \tag{2.41}$$

as well as the events

$$\mathcal{D}^k := \{\mathfrak{s}_k < \tau_B^\sigma\}, \qquad \forall k \in \mathbb{N}_0. \tag{2.42}$$

The cycle decomposition of $\{\sigma(t) : t \in [0, \tau_B^\sigma]\}$ is based on a collection of independent copies of Markov chains starting in η, $\{\eta^k = \{\eta^k(t) : t \in [0, \tau_B^{\eta,k}]\}\}_k$, and on a collection of i.i.d. stacks of coins $\{V^k \equiv \{V_0^k, \ldots, V_{M-1}^k\}\}$. The corresponding events $\{\mathcal{A}^k\}$ and $\{\mathcal{B}^k\}$ are well defined and independent.

To start with, we focus first of all on the case $\varrho^n(\sigma) = \varrho^n(\eta)$. In the first cycle we apply the coupling construction to the Markov chains $\{\sigma(t)\}$ starting in σ and $\{\eta^0(t)\}$ up to time T. On the event $\mathcal{A}^0 \cap \mathcal{B}^0$ we can update $\sigma(t)$ and $\eta(t)$ in such a way that $\sigma(t) = \eta(t)$ for all $t > T$. In this case the cycle decomposition terminates. On the event $(\mathcal{A}^0 \cap \mathcal{B}^0)^c$, we update $\sigma(t)$ and $\eta(t)$ independently for all $t > T$. If $\mathcal{D}^0$ occurs, the first cycle ends at the random time $\mathfrak{s}_0$ whereas if $(\mathcal{D}^0)^c$ happens, the cycle decomposition terminates as well, and a merging has not occurred.

Provided that the cycle decomposition has not terminated so far, in the $(k+1)$th cycle for $(k > 1)$, we consider the coupling $\{(\sigma(\mathfrak{s}_k + t), \eta^k(t))\}_{t=0}^T$ of the Markov chain $\{\sigma(t)\}$ starting from the configuration $\sigma(\mathfrak{s}_k)$ and an independent copy $\{\eta^k(t)\}$ of the η-chain starting in η using the independent sequence of coin tosses $\{V_1^k, \ldots, V_{M-1}^k\}$ in the coupling construction.

As a consequence, we arrive at the following decomposition of the hitting time τ_B^σ in terms of the (independent) hitting times $\tau_B^{\eta,k}$

$$\tau_B^\sigma = \sum_{k=0}^{\infty} \left(\mathfrak{s}_{k-1} + \tau_B^{\eta,k}\right) \mathbb{1}_{\mathcal{A}^k} \mathbb{1}_{\mathcal{B}^k} \prod_{l=0}^{k-1} \mathbb{1}_{\mathcal{D}^l} \left(1 - \mathbb{1}_{\mathcal{A}^l} \mathbb{1}_{\mathcal{B}^l}\right)$$

$$+ \sum_{k=0}^{\infty} \tau_B^\sigma \left(1 - \mathbb{1}_{\mathcal{D}^k}\right) \left(1 - \mathbb{1}_{\mathcal{A}^k} \mathbb{1}_{\mathcal{B}^k}\right) \prod_{l=0}^{k-1} \mathbb{1}_{\mathcal{D}^l} \left(1 - \mathbb{1}_{\mathcal{A}^l} \mathbb{1}_{\mathcal{B}^l}\right). \quad (2.43)$$

In the case $\varrho^n(\sigma) \neq \varrho^n(\eta)$ we cannot directly apply our coupling construction. Instead, we will first update $\sigma(t)$ and $\eta(t)$ independently until time $\mathfrak{s}_0$, provided that the event $\mathcal{D}^0$ occurs, and we use the cycle decomposition, as described above, afterwards. Let us emphasis that in the definition of $\mathfrak{s}_0$ the involved parameter T is equal to zero. This implies that we have to replace (2.43) by

$$\tau_B^\sigma = \tau_B^\sigma \left(1 - \mathbb{1}_{\mathcal{D}^0}\right) + \sum_{k=1}^{\infty} \left(\mathfrak{s}_{k-1} + \tau_B^{\eta,k}\right) \mathbb{1}_{\mathcal{A}^k} \mathbb{1}_{\mathcal{B}^k} \mathbb{1}_{\mathcal{D}^0} \prod_{l=1}^{k-1} \mathbb{1}_{\mathcal{D}^l} \left(1 - \mathbb{1}_{\mathcal{A}^l} \mathbb{1}_{\mathcal{B}^l}\right)$$

$$+ \sum_{k=1}^{\infty} \tau_B^\sigma \left(1 - \mathbb{1}_{\mathcal{D}^k}\right) \left(1 - \mathbb{1}_{\mathcal{A}^k} \mathbb{1}_{\mathcal{B}^k}\right) \mathbb{1}_{\mathcal{D}^0} \prod_{l=1}^{k-1} \mathbb{1}_{\mathcal{D}^l} \left(1 - \mathbb{1}_{\mathcal{A}^l} \mathbb{1}_{\mathcal{B}^l}\right). \quad (2.44)$$

In (2.43) and (2.44) we used the convention that products with a negative number of terms are equal to one.

Lemma 2.11. *Suppose that $A, B \subset \mathcal{S}_N$ mimic a metastable situation on level n. If n satisfies $\varepsilon(n) c_1 < (C - c)/2$ for some $c_1 > 1 + \alpha^{-1}$, then there exists a constant $c_7 > 0$ such that for all $\sigma, \eta \in A$ and N large enough*

$$\mathbb{P}_\sigma[\mathcal{D}^0] \geq 1 - e^{-c_7 N}. \quad (2.45)$$

where $\mathcal{D}^0$ depend on the configuration η.

Proof. Let us first consider the case when $\varrho^n(\sigma) = \varrho^n(\eta)$. To start with, fix $\delta = (C - c)/2$ and set $X = (\varrho^n)^{-1} \circ \varrho^n(\eta) \subset A$. Further, define

$$\boldsymbol{X}_\delta := \{\boldsymbol{z} \in \Gamma^n \mid \mathbb{P}_{\mu_X}[\tau_Z < \tau_X] > e^{-\delta N}, \, Z = \mathcal{S}^n[\boldsymbol{z}]\}$$

and set $X_\delta = \mathcal{S}^n[\boldsymbol{X}_\delta] \cup X$. Using the definition of $\mathcal{D}^0$ together with the strong Markov property we obtain that

$$\mathbb{P}_\sigma[\mathfrak{s}_0 < \tau_B] \geq \sum_{\xi \in X_\delta} \mathbb{P}_\sigma[\sigma(N^\kappa) = \xi] \, \mathbb{P}_\xi[\tau_X < \tau_B]$$

$$\geq \min_{\xi \in X_\delta} \mathbb{P}_\xi[\tau_X < \tau_B] \, \mathbb{P}_\sigma[\sigma(N^\kappa) \in X_\delta]. \quad (2.46)$$

2.4. Cycle decomposition of σ-paths

Hence, there remain two terms which we have to bound from below. We will first focus on the last term in (2.46). For any $Z = S^n[z]$ where $z \notin X_\delta$ it holds that $\mathbb{P}_{\mu_X}[\tau_Z < \tau_X] \leq e^{-\delta N}$. Moreover,

$$\mathbb{P}_\sigma[\sigma(N^\kappa) \in Z] \leq \mathbb{P}_\sigma[\tau_Z \leq N^\kappa] \leq N^\kappa \max_{\sigma \in X} \mathbb{P}_\sigma[\tau_Z < \tau_X], \qquad (2.47)$$

where we used in the last step that the Markov chain $\{\sigma(t)\}$ has to make one final excursion without returning to the starting set X and that there are at most N^κ attempts. On the other hand, Proposition 2.7 implies that

$$\mathbb{P}_\sigma[\tau_Z < \tau_X] \leq e^{\varepsilon(n)c_1 N} \mathbb{P}_{\mu_X}[\tau_Z < \tau_X] + e^{-c_2 N} \leq e^{-N(\delta - \varepsilon(n)c_1)} + e^{-c_2 N}. \qquad (2.48)$$

Hence, by combining the last two estimates we obtain

$$\mathbb{P}_\sigma[\sigma(N^\kappa) \in X_\delta] = 1 - \sum_{z \notin X_\delta} \mathbb{P}_\sigma[\sigma(N^\kappa) \in S^n[z]]$$

$$\geq 1 - N^{nq+\kappa}\left(e^{-N(\delta - \varepsilon(n)c_1)} + e^{-c_2 N}\right). \qquad (2.49)$$

Let us now focus on the first factor of the right-hand side of (2.46) for any arbitrary $\xi \in X_\delta$. Suppose that $\xi \in Z = S^n[z]$ with $z \in X_\delta$, then from (1.115) we have

$$\mathbb{P}_{\nu_{Z,X}}[\tau_B < \tau_X] = \frac{\text{cap}(B,X)}{\text{cap}(Z,X)} \mathbb{P}_{\nu_{B,X}}[\tau_Z < \tau_X] \leq \frac{\mathbb{P}_{\mu_X}[\tau_B < \tau_X]}{\mathbb{P}_{\mu_X}[\tau_Z < \tau_X]} \leq e^{-N(C-\delta)}, \qquad (2.50)$$

where we used (2.5) in the last step. Again, Proposition 2.7 implies that

$$\mathbb{P}_\xi[\tau_B < \tau_X] \leq e^{\varepsilon(n)c_1 N} \mathbb{P}_{\nu_{Z,X}}[\tau_B < \tau_X] + e^{-c_2 N} \leq e^{-N(C-\delta-\varepsilon(n)c_1)} + e^{-c_2 N}, \qquad (2.51)$$

whereas in the case when $\xi \in X$ we get that

$$\mathbb{P}_\xi[\tau_B < \tau_X] \leq e^{\varepsilon(n)c_1 N} \mathbb{P}_{\mu_X}[\tau_B < \tau_X] + e^{-c_2 N} \leq e^{-N(C-\varepsilon(n)c_1)} + e^{-c_2 N}. \qquad (2.52)$$

Thus, combining the last two estimates implies

$$\min_{\xi \in X_\delta} \mathbb{P}_\xi[\tau_X < \tau_B] \geq 1 - \left(e^{-N(C-\delta-\varepsilon(n)c_1)} + e^{-c_2 N}\right). \qquad (2.53)$$

It remains to consider the case that $\varrho^n(\sigma) \neq \varrho^n(\eta)$ for any $\sigma, \eta \in A$. Here, we denote by X, Y the set-valued preimage of $x = \varrho^n(\eta)$ and $y = \varrho^n(\sigma)$ under ϱ^n. Recall that in this case the event $\mathcal{D}^0$ is given by $\{\tau_X < \tau_B\}$. Hence, analog to (2.50) we have that

$$\mathbb{P}_{\nu_{Y,X}}[\tau_B < \tau_X] = \frac{\text{cap}(B,X)}{\text{cap}(Y,X)} \mathbb{P}_{\nu_{B,X}}[\tau_Y < \tau_X] \leq \frac{\mathbb{P}_{\mu_X}[\tau_B < \tau_X]}{\mathbb{P}_{\mu_X}[\tau_Y < \tau_X]} \leq e^{-N(C-c)}, \qquad (2.54)$$

where we used (2.6) in the last step. By Proposition 2.7 we obtain

$$\mathbb{P}_\sigma[\tau_B < \tau_X] \leq e^{\varepsilon(n)c_1 N} \mathbb{P}_{\nu_{Y,X}}[\tau_B < \tau_X] + e^{-c_2 N} \leq e^{-N(C-c-\varepsilon(n)c_1)} + e^{-c_2 N} \qquad (2.55)$$

which implies that

$$\mathbb{P}_\sigma[\tau_X < \tau_B] \geq 1 - \left(e^{-N(C-c-\varepsilon(n)c_1)} + e^{-c_2 N}\right). \qquad (2.56)$$

By assumption $\varepsilon(n)c_1 < (C-c)/2$. Hence, in view of (2.49), (2.53) and (2.56), there exists $c_7 > 0$ such that, for large enough N, (2.45) holds. □

2.5. Bounds on the mean hitting time and the Laplace transform

Finally, in a situation that mimics to be metastable, we prove the following pointwise estimate on metastable times.

Theorem 2.12. *Suppose that $A, B \subset \mathcal{S}_N$ mimics a metastable situation on level n, and assume that n satisfies $\varepsilon(n)c_1 < (C-c)/4$ for a given $c_1 > 1 + \alpha^{-1}$. Then, there exists $c_0 > 0$ such that, for N large enough,*

$$\max_{\sigma, \eta \in A} \left| \frac{\mathbb{E}_\sigma[\tau_B]}{\mathbb{E}_\eta[\tau_B]} - 1 \right| \leq e^{-c_0 N}. \qquad (2.57)$$

Proof. Let $\mathbb{E}_{\sigma,\eta}$ denote the expectation with respect to $\mathbb{P}_{\sigma,\eta}$ on the enlarged probability space. Provided that $\varrho^n(\sigma) = \varrho^n(\eta)$ for $\sigma, \eta \in A$, by (2.43) we have

$$\mathbb{E}_\sigma[\tau_B] \geq \sum_{k=0}^\infty \mathbb{E}_{\sigma,\eta}\left[\tau_B^{\eta,k} \mathbb{1}_{\mathcal{A}^k} \mathbb{1}_{\mathcal{B}^k} \prod_{l=0}^{k-1} \mathbb{1}_{\mathcal{D}^l}\left(1 - \mathbb{1}_{\mathcal{A}^l}\mathbb{1}_{\mathcal{B}^l}\right)\right] \qquad (2.58)$$

Further, let $\mathcal{F}_{\mathfrak{s}_k}$ denote the σ-algebra generated by all the events and trajectories $\mathcal{A}^l, \mathcal{B}^l, \mathcal{D}^l, \eta^l$ and $\{\sigma(t) : t \in (\mathfrak{s}_{l-1}, \mathfrak{s}_l]\}$ for $l \leq k$. Due to the independence of the copies $\{\eta^l, V^l\}$ it holds that

$$\mathbb{E}_{\sigma,\eta}\left[\tau_B^{\eta,k} \mathbb{1}_{\mathcal{A}^k}\mathbb{1}_{\mathcal{B}^k} \,\big|\, \mathcal{F}_{\mathfrak{s}_{k-1}}\right] = \mathbb{E}_\eta\left[\tau_B \mathbb{1}_A \mathbb{1}_B\right] = e^{-\varepsilon(n)M} \mathbb{E}_\eta\left[\tau_B \mathbb{1}_B\right] \qquad (2.59)$$

and for all $l < k$

$$\mathbb{E}_{\sigma,\eta}\left[\mathbb{1}_{\mathcal{D}^l}\left(1 - \mathbb{1}_{\mathcal{A}^l}\mathbb{1}_{\mathcal{B}^l}\right) \,\big|\, \mathcal{F}_{\mathfrak{s}_{l-1}}\right] \geq \mathbb{E}_\eta\left[1 - \mathbb{1}_A \mathbb{1}_B\right] - \max_{\sigma' \in A}\left(1 - \mathbb{P}_{\sigma'}[\mathcal{D}^0]\right)$$
$$\geq 1 - e^{-\varepsilon(n)M} - e^{-c_7 N} \qquad (2.60)$$

where we used (2.45) in the last step. By choosing $M = c_1 N$, as in Lemma 2.6, $T = N^\kappa$ for some $\kappa > 2$ and $c_1 > 1 + \alpha^{-1}$ large enough to ensure that $c_2 > 2c_5$, then

2.5. Bounds on the mean hitting time and the Laplace transform

(2.22) and (2.32) implies that

$$\mathbb{E}_\sigma[\tau_B] \geq \mathbb{E}_\eta[\tau_B \mathbb{1}_B] \, e^{-\varepsilon(n)c_1 N} \sum_{k=0}^\infty \left(1 - e^{-\varepsilon(n)c_1 N} - e^{-c_7 N}\right)^k$$

$$\geq \mathbb{E}_\eta[\tau_B] \frac{1 - e^{-c_4 N}}{1 + e^{-N(c_7 - \varepsilon(n)c_1)}}. \qquad (2.61)$$

If $\varrho^n(\sigma) \neq \varrho^n(\eta)$ for $\sigma, \eta \in A$, by using (2.44), an analog computation reveals that

$$\mathbb{E}_\sigma[\tau_B] \geq \mathbb{E}_\eta[\tau_B \mathbb{1}_B] \, e^{-\varepsilon(n)c_1 N} \left(1 - e^{-c_7 N}\right) \sum_{k=0}^\infty \left(1 - e^{-\varepsilon(n)c_1 N} - e^{-c_7 N}\right)^k$$

$$\geq \mathbb{E}_\eta[\tau_B] \frac{1 - e^{-c_4 N} - e^{-c_7 N}}{1 + e^{-N(c_7 - \varepsilon(n)c_1)}}. \qquad (2.62)$$

If $\varepsilon(n)c_1 < (C-c)/4$, we can choose c_7 in such a way that $c_7 > (C-c)/4$. Hence, there exists a constant $c_0 > 0$ such that $\mathbb{E}_\sigma[\tau_B] \geq \mathbb{E}_\eta[\tau_B](1 - e^{-c_0 N})$ for all $\sigma, \eta \in A$ and large enough N. This concludes the proof. □

As a further application of the coupling construction and the cycle decomposition we will prove a pointwise estimate on the Laplace transform of τ_B.

Proposition 2.13. *Suppose $A, B \subset S_N$ mimic a metastable situation on level n. If n satisfy $\varepsilon(n)c_1 < (C-c)/4$ for some $c_1 > 1 + \alpha^{-1}$ sufficiently large, then there exists $c_8 > 0$ such that for all $\lambda > 0$ and N large enough*

$$\max_{\sigma,\eta \in A} \left| \frac{\mathbb{E}_\sigma[e^{-\frac{\lambda}{T}\tau_B}]}{\mathbb{E}_\eta[e^{-\frac{\lambda}{T}\tau_B}]} - 1 \right| \leq e^{2\lambda} e^{-c_8 N}, \qquad (2.63)$$

where $T = \mathbb{E}_{\nu_A}[\tau_B]$ for some probability measure ν_A on A.

Proof. Let us point out that the proof of this proposition goes along the lines of [5, Prop. 3.4]. Analog to (2.43) we have, for any $\sigma, \eta \in A$ with $\varrho^n(\sigma) = \varrho^n(\eta)$,

$$e^{-\frac{\lambda}{T}\tau_B^\sigma} = \sum_{k=0}^\infty e^{-\frac{\lambda}{T}(s_{k-1} + \tau_B^{\eta,k})} \mathbb{1}_{A^k} \mathbb{1}_{B^k} \prod_{l=0}^{k-1} \mathbb{1}_{D^l}\left(1 - \mathbb{1}_{A^l}\mathbb{1}_{B^l}\right)$$

$$+ \sum_{k=0}^\infty e^{-\frac{\lambda}{T}\tau_B^\sigma} \left(1 - \mathbb{1}_{D^k}\right)\left(1 - \mathbb{1}_{A^k}\mathbb{1}_{B^k}\right) \prod_{l=0}^{k-1} \mathbb{1}_{D^l}\left(1 - \mathbb{1}_{A^l}\mathbb{1}_{B^l}\right).$$

This implies that

$$\mathbb{E}_\sigma\left[e^{-\frac{\lambda}{T}\tau_B}\right] \leq \sum_{k=0}^\infty \mathbb{E}_{\sigma,\eta}\left[e^{-\frac{\lambda}{T}\tau_B^{\eta,k}} \mathbb{1}_{A^k}\mathbb{1}_{B^k} \prod_{l=0}^{k-1}\mathbb{1}_{D^l}\left(1 - \mathbb{1}_{A^l}\mathbb{1}_{B^l}\right)\right]$$

$$+ \sum_{k=0}^\infty \mathbb{E}_{\sigma,\eta}\left[\left(1 - \mathbb{1}_{D^k}\right)\left(1 - \mathbb{1}_{A^k}\mathbb{1}_{B^k}\right) \prod_{l=0}^{k-1}\mathbb{1}_{D^l}\left(1 - \mathbb{1}_{A^l}\mathbb{1}_{B^l}\right)\right].$$

$$(2.64)$$

Analog to (2.59) and (2.60), we obtain, due to the independence of $\{\eta^l, V^l\}$,

$$\mathbb{E}_{\sigma,\eta}\left[\mathrm{e}^{-\frac{\lambda}{T}\tau_B^{\eta,k}}\mathbb{1}_{A^k}\mathbb{1}_{B^k}\,\big|\,\mathcal{F}_{s_{k-1}}\right] = \mathbb{E}_\eta\left[\mathrm{e}^{-\frac{\lambda}{T}\tau_B}\mathbb{1}_A\mathbb{1}_B\right] \leq \mathrm{e}^{-\varepsilon(n)M}\,\mathbb{E}_\eta\left[\mathrm{e}^{-\frac{\lambda}{T}\tau_B}\right] \quad (2.65)$$

and for all $l < k$

$$\mathbb{E}_{\sigma,\eta}\left[\mathbb{1}_{D^l}\left(1 - \mathbb{1}_{A^l}\mathbb{1}_{B^l}\right)\,\big|\,\mathcal{F}_{s_{l-1}}\right] \leq \mathbb{E}_\eta\left[1 - \mathbb{1}_A\mathbb{1}_B\right] = 1 - \mathrm{e}^{-\varepsilon(n)M}\,\mathbb{P}_\eta[\mathcal{B}]. \quad (2.66)$$

By choosing $M = c_1 N$, as in Lemma 2.6, and $T = N^\kappa$ for some $\kappa > 2$ and taking into account (2.45), we obtain for the second term in (2.64) the following upper bound

$$\mathrm{e}^{-c_7 N}\sum_{k=0}^{\infty}\left(1 - \mathrm{e}^{-\varepsilon(n)c_1 N}\left(1 - \mathrm{e}^{-c_4 N}\right)\right)^k \leq 2\,\mathrm{e}^{-N(c_7 - \varepsilon(n)c_1)}$$

for N large enough. Since $c_7 - \varepsilon(n)c_1 > (C - c)/4 - \varepsilon(n)c_1 > 0$ this upper bound is exponential small in N. Combining all the estimates above, we arrive at

$$\mathbb{E}_\sigma\left[\mathrm{e}^{-\frac{\lambda}{T}\tau_B}\right] \leq \mathbb{E}_\eta\left[\mathrm{e}^{-\frac{\lambda}{T}\tau_B}\right]\mathrm{e}^{-\varepsilon(n)c_1 N}\sum_{k=0}^{\infty}\left(1 - \mathrm{e}^{-\varepsilon(n)c_1 N}\left(1 - \mathrm{e}^{-c_4 N}\right)\right)^k$$
$$+ 2\,\mathrm{e}^{-N(c_7 - \varepsilon(n)c_1)}$$
$$\leq \mathbb{E}_\eta\left[\mathrm{e}^{-\frac{\lambda}{T}\tau_B}\right]\left(1 + 2\,\mathrm{e}^{-c_4 N}\right) + 2\,\mathrm{e}^{-N(c_7 - \varepsilon(n)c_1)}. \quad (2.67)$$

In the case when $\varrho^n(\sigma) = \varrho^n(\eta)$ for $\sigma, \eta \in A$, by a similar computation we obtain

$$\mathbb{E}_\sigma\left[\mathrm{e}^{-\frac{\lambda}{T}\tau_B}\right] \leq \mathbb{E}_\eta\left[\mathrm{e}^{-\frac{\lambda}{T}\tau_B}\right]\left(1 + 2\,\mathrm{e}^{-c_4 N}\right) + 2\,\mathrm{e}^{-N(c_7 - \varepsilon(n)c_1)} + \mathrm{e}^{-c_7 N}. \quad (2.68)$$

By Jensen's inequality and (2.57), for every $\eta \in A$ it holds that

$$\mathbb{E}_\eta\left[\mathrm{e}^{-\frac{\lambda}{T}\tau_B}\right] \geq \mathrm{e}^{-\frac{\lambda}{T}\mathbb{E}_\eta[\tau_B]} \geq \mathrm{e}^{-\lambda(1 + \mathrm{e}^{-c_0 N})} \geq \mathrm{e}^{-2\lambda}. \quad (2.69)$$

This implies that there exists $c_8 > 0$ such that $\mathbb{E}_\sigma\left[\mathrm{e}^{-\frac{\lambda}{T}\tau_B}\right] \leq \mathbb{E}_\eta\left[\mathrm{e}^{-\frac{\lambda}{T}\tau_B}\right]\left(1 + \mathrm{e}^{2\lambda - c_8 N}\right)$ for N large enough. $\square$

Chapter 3

Metastability

We turn now to a characterization of metastability for a class of reversible Markov chains. Our aim is to derive sharp estimates on metastable exit times both on the level of expected values and Laplace transforms by using the tools we have presented in the previous two chapters. For this reason, we restrict ourselves to the setting introduced in Chapter 2.

Starting point of our further investigations is the definition of a *set of metastable sets*. As a first step, we coarse-grain the state space by introducing a family of *mesoscopic variables* which depend on an additional parameter n controlling the coarsening level. The key idea behind our definition is that, for a fixed n, these sets are defined as preimages under the mesoscopic map of certain single points in the coarse-grained space. In doing so, the mesoscopic points are chosen in such a way to ensure that the process with initial distribution given by the reversible measure conditioned on one of the corresponding sets returns very often to it before escaping to a "more stable" set.

The present chapter is organized as follows. In Section 3.1 we start with reviewing some of the different definitions and approaches of metastability that have been introduced in the past. Afterwards, we present a definition of a set of metastable sets that will be used in the remaining part of this thesis. As an immediate consequence of this definition combined with a rough control on the regularity of harmonic functions, we show in Section 3.2 the existence of an ultra-metric on the set of metastable sets in the limit when N tends to infinity. As we will further show in Section 3.3, this property allows us to derive various bounds on harmonic functions and to prove sharp estimates for averaged mean hitting times. In the final Section 3.4 we prove the convergence of normalized metastable exit times to an exponential distribution. In the proof we use pointwise estimates of the Laplace transforms of metastable exit times as well as averaged renewal equations.

3.1. Introduction

In the diffusion setting, metastable sets are considerable easy to identify. Namely, they correspond to suitable chosen neighborhoods of local minima of the potential landscape. The dynamical aspects of metastability in such a context of randomly perturbed dynamical systems was first analyzed with mathematical rigor in the seminal work by Freidlin and Wentzell [51]. In models coming from statistical mechanics the question of how to characterize metastable states, and how to identify them in a given model context is really an issue.

Characterization of metastability: A partial review. Lebowitz and Penrose introduced a first dynamical approach to metastability that describes the metastable behavior of the systems in terms of *evolution of ensembles*. Their theory takes into account dynamical and statical properties of the system. They characterized *metastable thermodynamical states* by the following properties [96, 97]:

  (i) only one thermodynamical phase is present;
  (ii) if the system is isolated the lifetime of a metastable state is very large, i.e. a system that starts in this state is likely to take a long time to get out;
  (iii) the escape from a metastable state is an irreversible process, i.e. once the system has gotten out, it is unlikely to return.

Further, they invented the idea to define a metastable state by means of *restricted ensembles*, $\mu_X := \mu[\,\cdot\,|X]$, where μ is the equilibrium Gibbs measure of the system and X is a subset of the state space. The main task is to choose X is such a way that first μ_X describes a pure phase and secondly the escape rate from X at time zero is small. The latter implies that condition (ii) is satisfied provided that the escape rate is maximal at time zero. A necessary condition to satisfy the third criterion is to ensure that $\mu[X]$ is negligible. For an excellent review of this method we refer to the recent monograph by Olivieri and Vares [94].

Acting on the assumption that metastability is characterized by the existence of at least two different time scales for the evolution, Davies [29, 30, 31] suggested to analyze the small eigenvalues and to construct metastable sets from the corresponding eigenfunctions. Based on the observation that a Markov process, exhibiting a metastable behavior, is *almost reducible*, we can view the original, irreducible Markov process as a perturbation of a reducible one. Assuming that the process is reducible, the theorem of Perron-Frobenius implies that the generator has a degenerate eigenvalue

3.1. Introduction

zero whose multiplicity is equal to the number of different metastable sets. Moreover, the corresponding eigenfunctions are given as indicator functions on these sets. Provided that the perturbation is sufficiently small, this leads to a cluster of small eigenvalues that is separated by a gap from the rest of the spectrum. This approach has been continued by Gaveau and Schulman [55] and more recently by Huisinga, Meyn and Schütte [66]. In particular, in the study of metastable chemical conformations of biomolecules Deuflhard et al. introduced a numerical algorithm to determine metastable sets [36, 37].

From the theoretical point of view, the spectral signature of metastability is maybe the most elegant and complete way to characterize this phenomenon in Markov processes. However, in the majority of applications coming from statistical mechanics it is very hard to determine both the small eigenvalues and the corresponding eigenvectors explicitly.

In a recent paper [7], Bianchi and Gaudillière presented an approach to metastability that combines aspects of the idea of the evolution of ensembles and spectral properties. Instead of considering the restricted ensemble μ_X, they suggested to characterize metastable states by the quasi-stationary measure μ_X^* that is the left eigenvector to the principle Dirichlet eigenvalue of the generator subject to Dirichlet boundary conditions on X^c. The advantage of choosing μ_X^* as the initial distribution of the Markov process is that the exit times of the set X are exponential distributed. Provided that the ratio between the principle Dirichlet eigenvalue and the principle eigenvalue with respect to the reflected process can be controlled, they established sharp estimates on the mean exit times.

As it was pointed out in [23], the approach of evolution of ensembles has the drawback that by studying the time evolution of probability distributions it is difficult to distinguish between a smooth but very slow relaxation towards equilibrium and a typical metastable behavior. The latter is characterized by the fact that the process is apparently stationary for a long period until it undergoes a rapid transition at some randomly distributed point in time leading the system to a different state that again seems be to stationary. Based on this fact Cassandro, Galves, Olivieri and Vares proposed a *pathwise approach to metastability* that uses time averages along trajectories to single out properties of typical paths as well as their statistics [23, 94]. The main tools in their approach are large deviation methods in path space and model reduction techniques. The latter, originally introduced by Freidlin and Wentzell in the context of randomly perturbed dynamical systems, allows to reduce the problem of describing the long-time behavior of a process in terms of a *Markov chain with exponentially small transition probabilities* between metastable sets. This approach has been proven to be robust and rather universally applicable in various model contexts.

While the pathwise approach to metastability is able to yield detailed informations, for instance on the typical exit path, its precision to predict, e.g. the mean exit times, is limited to logarithmic equivalence.

More recently, Bovier, Gayrard, Eckhoff and Klein introduced a *potential theoretic approach to metastability* [15, 16] that was systematically developed in the sequel. Their starting point is the definition of a set of *metastable points*, $\mathcal{M}$, that is characterized by the ratio of two *escape probabilities*. More precisely, for any $\sigma \in \mathcal{M}$ the probability to escape from σ to the remaining metastable points $\mathcal{M} \setminus \{\sigma\}$ should be much smaller compared to the probability to reach $\mathcal{M}$ starting from some arbitrary point in the state space outside $\mathcal{M}$ before returning to it. The key idea of this approach is to express quantities of interest in terms of capacities and use variational principles to compute the latter. Having identified the set of metastable points, this approach allows to prove sharp estimates on mean exit times, to analyze precisely the low-lying spectrum of the corresponding generator and to control the deviation of the law of metastable exit times from the exponential distribution for a wide class of *reversible* Markov processes. Thereby, strict renewal equations play an important role to derive various estimates. For an introduction to the potential theoretic approach to metastability, we refer to [11].

As it was already pointed out in [9], the definition of metastable points relies crucially on the fact that the time to reach the set of metastable points is small compared to the transition time between different elements of this set. However, in the context of reversible diffusion processes or spin systems at finite temperature the probability to hit a given point in finite time is either zero or exponentially small. For diffusion processes, this problem can be overcome by considering around each metastable point, x, a small ball, $B_\varepsilon(x)$, of radius $\varepsilon > 0$. In this spirit, a definition of a set of metastable points was introduced in [10]. Based on Harnack and Hölder inequalities, apriori estimates on the local regularity of solutions of boundary value problems with respect to the generator of the diffusion process yield that the typical oscillations of such functions within $B_\varepsilon(x)$ are bounded by some polynomial in ε. Provided that ε is chosen appropriately, this allows to establish strict renewal equations e.g. for harmonic functions or Laplace transforms of mean hitting times [17, 18].

Metastable sets and local valleys. Our starting point in the study of the dynamical behavior of the Markov chain $\{\sigma(t)\}$ is the following definition of a *set of metastable sets*.

Definition 3.1. Let $\mathcal{M}^n \equiv \mathcal{M}_N^n$ be a set of disjoint subsets of $\mathcal{S}_N$. Assume that there exists $n \in \mathbb{N}$ such that $n \ll N$ and for every $A \in \mathcal{M}^n$ there exists $a \in \Gamma^n$ such that A is given as the set-valued preimage of a under ϱ^n, i.e. $A = \mathcal{S}^n[a]$. Further, let $\mathcal{S}_N^n := \{\mathcal{S}^n[x] \subset \mathcal{S}_N \mid x \in \Gamma^n\}$. Then, $\mathcal{M}^n$ is called a *set of metastable sets* if there exists $\mathfrak{C} \equiv \mathfrak{C}(n) > 0$ such that

$$\frac{\max_{A \in \mathcal{M}^n} \mathbb{P}_{\mu_A}[\tau_{\mathcal{M}^n \setminus A} < \tau_A]}{\min_{X \in \mathcal{S}_N^n \setminus \mathcal{M}^n} \mathbb{P}_{\mu_X}[\tau_{\mathcal{M}^n} < \tau_X]} \leq e^{-\mathfrak{C}N}. \tag{3.1}$$

Here, $\mu_X(\sigma) = \mu_N[\sigma | X]$ for $\sigma \in X \subset \mathcal{S}_N$ stands for the reversible measure, μ_N, conditioned on the set X.

Remark 3.2. The definition above is a natural generalization of the one for metastable points given in [**9, 11**].

Remark 3.3. Depending on the choice of $\mathfrak{C}$, we may find different sets $\mathcal{M}^n$. Let us emphasize the fact that, in the considered setting, the cardinality of $\mathcal{M}^n$ grows at most polynomial with N. Moreover, the equation (3.1) has the pleasant feature that in the choice of the elements of $\mathcal{M}^n$ we have to take into account only sufficiently "deep minima". This allow us to deal with situations where an exponential large number (in N) of "shallow local minima" are present.

Remark 3.4. In order to compute precisely e.g. metastable exit times, in the approach presented below we need as an input some estimates of the oscillations of harmonic functions within mesoscopic sets. However, in view of (3.1), it suffice to control these oscillations on exponential scales.

We associate to each $M \in \mathcal{M}^n$ a *local valley on mesoscopic level* n

$$\mathcal{V}^n(M) := \left\{ X \subset \mathcal{S}_N \mid \exists \boldsymbol{x} \in \Gamma^n : X = \mathcal{S}^n[\boldsymbol{x}], \, \mathbb{P}_{\mu_X}[\tau_M \leq \tau_X] \geq \mathbb{P}_{\mu_X}[\tau_{\mathcal{M}^n \setminus M} \leq \tau_X] \right\}.$$

Notice that the sets $\mathcal{V}^n(M)$ are not necessary disjoint. By using arguments similar to the ones given in [**16**] in the case of points, we will show, see Corollary 3.7, that the set of sets which belong to more than one local valley has a very small mass under μ_N. Since the conditions above do not uniquely determine $\mathcal{M}^n$, it will be reasonable to choose $\mathcal{M}^n$ in such a way that for all $M \in \mathcal{M}^n$

$$\mu_N[X] \leq \mu_N[M], \qquad \forall X \in \mathcal{V}^n(M). \tag{3.2}$$

Furthermore, for every $A \in \mathcal{M}^n$ we denote by $\mathcal{M}^n(A) := \{X \in \mathcal{M}^n \mid \mu_N[X] > \mu_N[A]\}$ the subset of "deeper" metastable sets.

3.2. Ultrametricity

Let us start with studying the capacity between different metastable sets. More precisely, for an arbitrary $M \in \mathcal{M}^n$ and $B \subset \mathcal{M}^n \setminus M$ our aim is to show that if $X \in \mathcal{V}^n(M)$ then either the escape probability $\mathbb{P}_{\mu_X}[\tau_B < \tau_X]$ can be bounded from below, or the capacity $\text{cap}(X, B)$ is essentially the same as $\text{cap}(M, B)$. The strategy to prove such a statement relies on the ultrametricity of capacities, as specified in Lemma 1.26. In order to verify the assumptions given there, we would like to take advantage of Proposition 2.7. However, the bounds obtained on harmonic functions are unpleasant to derive a suitable upper bound on (1.112). For this reason, our starting point in the proof is instead of (1.113) the equation (1.115).

Lemma 3.5. *Let* $M \in \mathcal{M}^n$, $B \subset \mathcal{M}^n \setminus M$ *and* $X \in \mathcal{V}^n(M) \setminus M$. *Further, let* $c_1 > 1 + \alpha^{-1}$ *and choose* $\delta_1 > \varepsilon(n) c_1$. *Suppose that* $\mathfrak{C} - \delta_1 > 0$, *then either*

$$\mathbb{P}_{\mu_X}\left[\tau_B < \tau_X\right] \geq \frac{1}{2} e^{N(\mathfrak{C}-\delta_1)} \max_{A \in \mathcal{M}^n} \mathbb{P}_{\mu_A}\left[\tau_{\mathcal{M}^n \setminus A} < \tau_A\right] \qquad (3.3)$$

or

$$1 - e^{-N(\delta_1 - \varepsilon(n) c_1)} - e^{-c_2 N} \leq \frac{\mathrm{cap}(X, B)}{\mathrm{cap}(M, B)} \leq \left(1 - e^{-N(\delta_1 - \varepsilon(n) c_1)} - e^{-c_2 N}\right)^{-1}. \qquad (3.4)$$

Remark 3.6. *Since* $\varepsilon(n)$ *is decreasing in* n, *by choosing* n *in the definition of metastable sets large enough, we can ensure that* $\mathfrak{C} - \delta_1 > 0$.

Proof. First of all note that the definition of a local valley on mesoscopic level n implies that for all $X \in \mathcal{V}^n(M)$

$$\mathbb{P}_{\mu_X}\left[\tau_{\mathcal{M}^n} < \tau_X\right] \leq \mathbb{P}_{\mu_X}\left[\tau_M < \tau_X\right] + \mathbb{P}_{\mu_X}\left[\tau_{\mathcal{M}^n \setminus M} < \tau_X\right] \leq 2\,\mathbb{P}_{\mu_X}\left[\tau_M < \tau_X\right]. \qquad (3.5)$$

Hence, together with the definition of metastable sets, see (3.1), we get that

$$\max_{A \in \mathcal{M}^n} \mathbb{P}_{\mu_A}\left[\tau_{\mathcal{M}^n \setminus A} < \tau_A\right] \leq e^{-\mathfrak{C}N} \mathbb{P}_{\mu_X}\left[\tau_{\mathcal{M}^n} < \tau_X\right] \leq 2 e^{-\mathfrak{C}N} \mathbb{P}_{\mu_X}\left[\tau_M < \tau_X\right] \qquad (3.6)$$

Let us now assume that $\mathrm{cap}(X, B) > e^{-\delta_1 N} \mathrm{cap}(X, M)$. Then, the combination of the two equations above yields immediately (3.3). Hence, we are left with considering the remaining case where $\mathrm{cap}(X, B) \leq e^{-\delta_1 N} \mathrm{cap}(X, M)$. Here, our starting point is (1.115)

$$1 - \mathbb{P}_{\nu_{M,B}}\left[\tau_B < \tau_X\right] \leq \frac{\mathrm{cap}(X, B)}{\mathrm{cap}(M, B)} \leq \left(1 - \mathbb{P}_{\nu_{X,B}}\left[\tau_B < \tau_M\right]\right)^{-1}. \qquad (3.7)$$

Our strategy is the following: with the help of Proposition 2.7 we replace the initial distribution, $\nu_{M,B}$, on the left-hand side of (3.7) by the distribution $\nu_{M,X}$ which has the advantage that by a computation analog to (1.115) combined with the assumption $\mathrm{cap}(X, B) \leq e^{-\delta_1 N} \mathrm{cap}(X, M)$ we obtain immediately that

$$\mathbb{P}_{\nu_{M,X}}\left[\tau_B < \tau_X\right] \leq \frac{\mathrm{cap}(B, X)}{\mathrm{cap}(M, X)} \leq e^{-\delta_1 N}. \qquad (3.8)$$

On the right-hand side of (3.7) we proceed similarly and replace $\nu_{X,B}$ by $\nu_{X,B \cup M}$. Then, an application of the averaged renewal equation for capacities (1.89) reveals

3.2. Ultrametricity

that
$$\mathbb{P}_{\nu_{X,B\cup M}}[\tau_B < \tau_M] \leq \frac{\mathrm{cap}(X,B)}{\mathrm{cap}(X,M)} \leq e^{-\delta_1 N}.$$

Let us now describe how the actual replacement of the initial distributions is done. Since M, B as well as X are given as preimages under ϱ^n of certain points in Γ^n, Proposition 2.7 is applicable. Hence, for all $\sigma, \eta \in M$,

$$\mathbb{P}_\sigma[\tau_B < \tau_X] \leq e^{\varepsilon(n)c_1 N} \, \mathbb{P}_\eta[\tau_B < \tau_X] + e^{-c_2 N}. \tag{3.9}$$

By multiplying both sides with $\nu_{M,X}(\eta)$ and summing over all $\eta \in M$, we obtain the following upper bound

$$\mathbb{P}_\sigma[\tau_B < \tau_X] \leq e^{\varepsilon(n)c_1 N} \, \mathbb{P}_{\nu_{M,X}}[\tau_B < \tau_X] + e^{-c_2 N}$$
$$\leq e^{-N(\delta_1 - \varepsilon(n)c_1)} + e^{-c_2 N}, \tag{3.10}$$

uniformly for all $\sigma \in M$. By a similar computation we have that

$$\mathbb{P}_\sigma[\tau_B < \tau_M] \leq e^{\varepsilon(n)c_1 N} \, \mathbb{P}_{\nu_{X,B\cup M}}[\tau_B < \tau_M] + e^{-c_2 N}$$
$$\leq e^{-N(\delta_1 - \varepsilon(n)c_1)} + e^{-c_2 N}, \tag{3.11}$$

uniformly for all $\sigma \in X$. Hence, by combining (3.10) and (3.11) with (3.7) the estimate (3.4) follows. □

A simple corollary of Lemma 3.5 shows that sets $X = S^n[x]$ that belong to more than one local valley, $\mathcal{V}^n(M)$, have a vanishing mass under the invariant measure μ_N.

Corollary 3.7. *Consider two distinct metastable sets $A, B \in \mathcal{M}^n$ with $\mathcal{V}^n(A) \cap \mathcal{V}^n(B) \neq \emptyset$. Further, let $c_1 > 1 + \alpha^{-1}$ and $\delta_1 > \varepsilon(n)c_1$. If $\mathfrak{C} - \delta_1 > 0$, then for all $X \in \mathcal{V}^n(A) \cap \mathcal{V}^n(B)$*

$$\mu_N[X] \leq 2 e^{-N(\mathfrak{C} - \delta_1)} \min\{\mu_N[A], \mu_N[B]\}. \tag{3.12}$$

Proof. Suppose that $\mu_N[B] < \mu_N[A]$. Since $X \in \mathcal{V}^n(A) \cap \mathcal{V}^n(B)$, (3.6) implies that

$$\mathbb{P}_{\mu_A}[\tau_B < \tau_A] \leq 2 e^{-\mathfrak{C} N} \, \mathbb{P}_{\mu_X}[\tau_A < \tau_X] \tag{3.13}$$

and

$$\mathbb{P}_{\mu_A}[\tau_B < \tau_A] \leq 2 e^{-\mathfrak{C} N} \, \mathbb{P}_{\mu_X}[\tau_B < \tau_X]. \tag{3.14}$$

In particular, we have that

$$\mathrm{cap}(A,B) \leq 2 e^{-\mathfrak{C} N} \frac{\mu_N[A]}{\mu_N[X]} \mathrm{cap}(B,X).$$

In the remaining part of the proof we have to consider two different cases. Let us first of all assume that $2 e^{-\mathfrak{C} N} \mu_N[A]/\mu_N[X] \leq e^{-\delta_1 N}$. Then, by Lemma 3.5, we obtain that

$$\mathrm{cap}(A,X) \leq \left(1 - e^{-N(\delta_1 - \varepsilon(n)c_1)} - e^{-c_2 N}\right)^{-1} \mathrm{cap}(A,B). \tag{3.15}$$

Hence,

$$\mu_N[X] \le \left(1 - e^{-N(\delta_1 - \varepsilon(n)c_1)} - e^{-c_2 N}\right)^{-1} \frac{\mathbb{P}_{\mu_A}[\tau_B < \tau_A]}{\mathbb{P}_{\mu_X}[\tau_A < \tau_X]} \mu_N[A]$$

$$\le 2\left(1 - e^{-N(\delta_1 - \varepsilon(n)c_1)} - e^{-c_2 N}\right)^{-1} e^{-\mathfrak{C}N} \mu_N[A], \tag{3.16}$$

where we used in the first step that $\operatorname{cap}(A, X) = \mu_N[X] \mathbb{P}_{\mu_X}[\tau_A < \tau_X]$, while in the second step we took advantage of (3.13) and (3.14). On the other hand, if $2 e^{-\mathfrak{C}N} \mu_N[A]/\mu_N[X] > e^{-\delta_1 N}$ the assertion of the corollary follows easily. □

3.3. Sharp estimates on mean hitting times

In this section we demonstrate how the averaged renewal equation (1.89) and the almost ultrametricity of capacities (1.113) combined with the coupling method can be used to prove various estimates on harmonic functions. In particular, we derive a precise formula for mean hitting times of metastable sets. The control obtained here is crucial for the investigation of the distribution of metastable exit times in Section 3.4.

Let $\mathcal{M}^n$ be a set of metastable sets and $\mathfrak{C}$ as defined in Definition 3.1. Let us now choose some $A \in \mathcal{M}^n$ and let $B \subset \mathcal{M}^n(A)$. Moreover, we may rewrite

$$\mathbb{E}_{\nu_{A,B}}[\tau_B] = \sum_{\sigma \in A} \nu_{A,B}(\sigma)\, \mathbb{E}_{\sigma}[\tau_B] = \frac{1}{\operatorname{cap}(A, B)} \sum_{x \in \Gamma^n} \sum_{\sigma \in S^n[x]} \mu_N(\sigma)\, h_{A,B}(\sigma).$$
(3.17)

Our aim is to show that, under a non-degeneracy condition that will be specified below, the sum on the right-hand side of (3.17) is of the order $\mu_N[A]$.

The following lemma provides a necessary control over the equilibrium potential, $h_{A,B}$, within the different local valleys.

Lemma 3.8. *Let $A \in \mathcal{M}^n$ and $B \subset \mathcal{M}^n \setminus A$. Further, suppose that $c_1 > 1 + \alpha^{-1}$ is chosen in such a way that $e^{-c_2 N} < \mathbb{P}_{\mu_A}[\tau_B < \tau_A]$. Then,*

(i) *for every $M \in B$ and $X \in \mathcal{V}^n(M) \setminus M$*

$$\sum_{\sigma \in X} \mu_N(\sigma)\, h_{A,B}(\sigma) \le 2 e^{-N(\mathfrak{C} - \varepsilon(n)c_1)} \left(1 - 2 e^{-\mathfrak{C}N}\right)^{-1} \mu_N[A], \tag{3.18}$$

(ii) *for $X \in \mathcal{V}^n(A) \setminus A$*

$$\sum_{\sigma \in X} \mu_N(\sigma)\left(1 - h_{A,B}(\sigma)\right) \le 2 e^{-N(\mathfrak{C} - \varepsilon(n)c_1)} \left(1 - 2 e^{-\mathfrak{C}N}\right)^{-1} \mu_N[A], \tag{3.19}$$

(iii) *for every $M \in \mathcal{M}^n \setminus (\{A\} \cup B)$ and $X \in \mathcal{V}^n(M)$ either*

$$\sum_{\sigma \in X} \mu_N(\sigma)\, h_{A,B}(\sigma) \le 2 e^{-N(\mathfrak{C} - \delta_1 - \varepsilon(n)c_1)} \left(1 - 2 e^{-N(\mathfrak{C} - \delta_1)}\right)^{-1} \mu_N[A], \tag{3.20}$$

3.3. Sharp estimates on mean hitting times

or

$$\sum_{\sigma \in X} \mu_N(\sigma)\, h_{A,B}(\sigma)$$
$$\leq a(n,N)\, e^{\varepsilon(n)c_1 N} \left(1 - a(n,N)\, \frac{e^{-c_2 N}}{\mathbb{P}_{\mu_M}[\tau_B < \tau_M]}\right)^{-1} \frac{\mathbb{P}_{\mu_A}[\tau_B < \tau_A]}{\mathbb{P}_{\mu_M}[\tau_B < \tau_M]}\, \mu_N[A], \tag{3.21}$$

where $a(n,N)^{-1} = 1 - 2\,e^{-N(\delta_1 - \varepsilon(n)c_1)}$ and $0 < \delta_1 < \mathfrak{C}$. Moreover, the same statement holds for $1 - h_{A,B}$ except that the term $\mathbb{P}_{\mu_M}[\tau_B < \tau_M]$ in (3.21) has to be replaced by the probability $\mathbb{P}_{\mu_M}[\tau_A < \tau_M]$.

Proof. The idea behind the proof is based on the representation (1.115) and pointwise estimates on harmonic functions (2.27). In view of the definition of metastable sets (3.1), we obtain first of all that $\mathbb{P}_{\mu_A}[\tau_B < \tau_A] \leq e^{-\mathfrak{C}N}\, \mathbb{P}_{\mu_X}[\tau_{\mathcal{M}^n} < \tau_X]$. Hence, for every $M \in \mathcal{M}^n$ and $X \in \mathcal{V}^n(M) \setminus M$ it holds that

$$\mathbb{P}_{\mu_A}[\tau_B < \tau_A] \leq 2\,e^{-\mathfrak{C}N}\, \mathbb{P}_{\mu_X}[\tau_M < \tau_X] \tag{3.22}$$

where we used (3.6).

(i) and (ii). Due to the fact that the proof of (3.18) and (3.19) are completely similar, we will give a detailed proof only for (i). Let us fix some $M \in B$. Since $h_{A,B}(\sigma) = 0$ for all $\sigma \in B$, it remains to consider all $X \in \mathcal{V}^n(M) \setminus M$. Recall that, by definition, the metastable sets are preimages under ϱ^n of some points in Γ^n. Hence, in view of (2.27), we have that

$$\mathbb{P}_\sigma[\tau_B < \tau_X] \geq e^{-\varepsilon(n)c_1 N} \left(\mathbb{P}_\eta[\tau_B < \tau_X] - e^{-c_2 N}\right) \tag{3.23}$$

for all $\sigma, \eta \in X$. By multiplying both sides with $\mu_N(\eta)$ and summing over all $\eta \in X$, we obtain the following pointwise estimate

$$\mathbb{P}_\sigma[\tau_B < \tau_X] \geq e^{-\varepsilon(n)c_1 N}\, \frac{\mathrm{cap}(B,X)}{\mu_N[X]} \left(1 - \frac{e^{-c_2 N}}{\mathbb{P}_{\mu_X}[\tau_B < \tau_X]}\right). \tag{3.24}$$

On the other hand, a computation analog to (1.115) shows that

$$\mathbb{P}_{\nu_{X,B}}[\tau_A < \tau_B] \leq \frac{\mathrm{cap}(A,B)}{\mathrm{cap}(X,B)} = \frac{\mathbb{P}_{\mu_A}[\tau_B < \tau_A]}{\mathbb{P}_{\mu_X}[\tau_B < \tau_X]}\, \frac{\mu_N[A]}{\mu_N[X]}. \tag{3.25}$$

By using (3.24), we can replace the last exit biased distribution on the left-hand side of (3.25) through the conditional reversible measure μ_X. Thus,

$$\sum_{\sigma \in X} \mu_N(\sigma)\, h_{A,B}(\sigma) \leq e^{\varepsilon(n)c_1 N} \left(1 - \frac{e^{-c_2 N}}{\mathbb{P}_{\mu_X}[\tau_B < \tau_X]}\right)^{-1} \frac{\mathbb{P}_{\mu_A}[\tau_B < \tau_A]}{\mathbb{P}_{\mu_X}[\tau_B < \tau_X]}\, \mu_N[A]$$
$$\leq 2\,e^{-N(\mathfrak{C} - \varepsilon(n)c_1)} \left(1 - 2\,e^{-\mathfrak{C}N}\right)^{-1} \mu_N[A], \tag{3.26}$$

where we used in the last step that $\mathbb{P}_{\mu_A}[\tau_B < \tau_A] \leq 2\,e^{-\mathfrak{C}N}\,\mathbb{P}_{\mu_X}[\tau_B < \tau_X]$ as well as the assumption on the choice of c_2. This concludes the proof of (i) and (ii).

(iii) Let $X \in \mathcal{V}^n(M)$ where $M \in \mathcal{M}^n \setminus \{A\} \cup B$. If $X \neq M$, an application of (3.3) for some $0 < \delta_1 < \mathfrak{C}$ reveals that $\mathbb{P}_{\mu_A}[\tau_B < \tau_A] \leq 2\,e^{-N(\mathfrak{C}-\delta_1)}\,\mathbb{P}_{\mu_X}[\tau_B < \tau_X]$. Hence, by (3.26),

$$\sum_{\sigma \in X} \mu_N(\sigma)\,h_{A,B}(\sigma) \leq 2\,e^{-N(\mathfrak{C}-\delta_1-\varepsilon(n)c_1)}\left(1 - 2\,e^{-N(\mathfrak{C}-\delta_1)}\right)^{-1} \mu_N[A]. \quad (3.27)$$

On the other hand, due to (3.4), we can bound the probability $\mathbb{P}_{\mu_X}[\tau_B < \tau_X]$ from below by $a(n,N)^{-1}\,\mathbb{P}_{\mu_M}[\tau_B < \tau_M]$. By plugging this estimate into (3.26), we obtain (3.21) for $X \in \mathcal{V}^n(M) \setminus M$. Notice that (3.21) follows immediately from the equation (3.26) in the case when $X = M$. The completes the proof. $\square$

In view of Lemma 3.8, we can derive precise expressions for mean hitting times (3.17) that involve only capacities and the invariant measure of local valleys. For simplicity, let us take into account the following additional *non-degeneracy* condition:

Definition 3.9. Let $A \in \mathcal{M}^n$ and $B \subset \mathcal{M}^n \setminus A$. We say that the non-degeneracy condition is satisfied, if there exists a $0 < \delta < \mathfrak{C}$ such that either

$$\mu_N[M] < e^{-\delta N}\,\mu_N[A] \quad \text{or} \quad \mathbb{P}_{\mu_A}[\tau_B < \tau_A] < e^{-\delta N}\,\mathbb{P}_{\mu_M}[\tau_B < \tau_M] \quad (3.28)$$

for all $M \in \mathcal{M}^n \setminus (A \cup B)$.

We can now easily prove the following

Theorem 3.10. *For $A \in \mathcal{M}^n$ and $B \subset \mathcal{M}^n(A)$ assume that the non-degeneracy condition for some $0 < \delta < \mathfrak{C}$ is satisfied. If $n \ll N$ is such that $\varepsilon(n)c_1 < \min\{\delta, \mathfrak{C} - \delta\}$ where c_1 is chosen in such a way that $e^{-c_2 N} < \mathbb{P}_{\mu_A}[\tau_B < \tau_A]$, then there exists $c_9 > 0$ such that for N large enough*

$$\mathbb{E}_{\nu_{A,B}}[\tau_B] = \frac{\mu_N[\mathcal{V}^n(A)]}{\mathrm{cap}(A,B)}\left(1 + e^{-c_9 N}\right). \quad (3.29)$$

Proof. For δ and n as specified above, let us define

$$\mathcal{U}_\delta^n := \left\{ X \subset \mathcal{S}_N \;\middle|\; \exists\, x \in \Gamma^n : X = \mathcal{S}^n[x],\; \mu_N[X] \geq \mu_N[A]\,e^{-\delta N} \right\}.$$

An obvious consequence is the following

Lemma 3.11. *With the notations introduced above, it holds that*

$$\sum_{X \notin \mathcal{U}_\delta^n} \sum_{\sigma \in X} \mu_N(\sigma)\,h_{A,B}(\sigma) \leq N^{nq}\,e^{-\delta N}\,\mu_N[A]. \quad (3.30)$$

3.3. Sharp estimates on mean hitting times

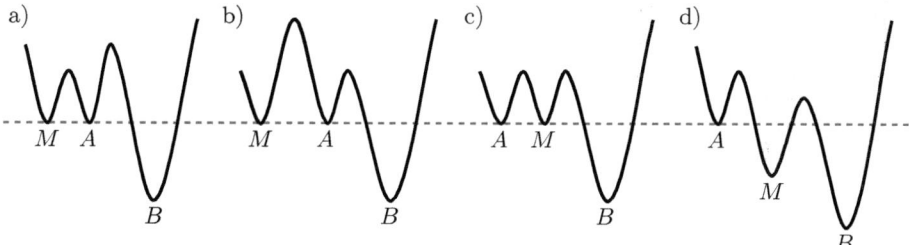

Figure 1. Illustration of various different landscapes that are excluded by the non-degeneracy condition.

It remains to control the harmonic function $h_{A,B}$ for all configurations $\sigma \in X$ where $X \in \mathcal{U}_\delta$. For this purpose, let us rewrite the neighborhood $\mathcal{U}_\delta^n$ as

$$\mathcal{U}_\delta^n = \bigcup_{M \in \mathcal{M}^n} \mathcal{U}_\delta^n(M)$$

where $\mathcal{U}_\delta^n(M) := \mathcal{U}_\delta^n \cap \mathcal{V}^n(M)$ for $M \in \mathcal{M}^n$. Notice that for $M, M' \in \mathcal{M}^n$ it may happen that $\mathcal{U}_\delta^n(M) \cap \mathcal{U}_\delta^n(M') \neq \emptyset$. We begin with controlling $h_{A,B}$ on $\mathcal{U}_\delta^n(M)$ for all $M \in \{A\} \cup B$. In view of Lemma 3.8 it is immediate that, for N large enough,

$$\sum_{X \in \mathcal{U}_\delta^n(A)} \sum_{\sigma \in X} \mu_N(\sigma) \left(1 - h_{A,B}(\sigma)\right) \leq N^{nq} e^{-N(\mathfrak{C}-\varepsilon(n)c_1)} \mu_N[A] \quad (3.31)$$

while for all $M \in B$

$$\sum_{X \in \mathcal{U}_\delta^n(A)} \sum_{\sigma \in X} \mu_N(\sigma) \left(1 - h_{A,B}(\sigma)\right) \leq N^{nq} e^{-N(\mathfrak{C}-\varepsilon(n)c_1)} \mu_N[A]. \quad (3.32)$$

As a next step, let us consider the contribution to (3.17) coming from the neighborhood $\mathcal{U}_\delta^n(M)$ for $M \in \mathcal{M}^n \setminus (\{A\} \cup B)$. For this purpose, we set $\delta_1 = \delta$ in Lemma 3.8 (iii). Then, in the case when $\operatorname{cap}(X, B) > e^{-\delta N} \operatorname{cap}(X, M)$, (3.20) implies that

$$\sum_{X \in \mathcal{U}_\delta^n(M)} \sum_{\sigma \in X} \mu_N(\sigma) h_{A,B}(\sigma) \leq N^{nq} e^{-N(\mathfrak{C}-\delta-\varepsilon(n)c_1)} \mu_N[A]. \quad (3.33)$$

Hence, it remains to consider the case $\operatorname{cap}(X, B) \leq e^{-\delta N} \operatorname{cap}(X, M)$. As a consequence of the non-degeneracy condition, $\mathbb{P}_{\mu_A}\left[\tau_B < \tau_A\right] \leq e^{-\delta N} \mathbb{P}_{\mu_M}\left[\tau_B < \tau_M\right]$. On the other hand, by assumption, it holds that $e^{-c_2 N} < \mathbb{P}_{\mu_A}\left[\tau_B < \tau_A\right]$. Therefore, in view of (3.21),

$$\sum_{X \in \mathcal{U}_\delta^n(M)} \sum_{\sigma \in X} \mu_N(\sigma) h_{A,B}(\sigma) \leq N^{nq} e^{-N(\delta-\varepsilon(n)c_1)} \mu_N[A]. \quad (3.34)$$

Now, the proof of (3.29) is an immediate consequence of (3.17) combined with (3.30) and (3.32) – (3.34). Namely, equation (3.30) together with (3.32), (3.33) an (3.34) provide the following upper bound

$$\sum_{\sigma \in S_N} \mu_N(\sigma) h_{A,B}(\sigma)$$

$$\leq \sum_{X \in \mathcal{V}^n(A)} \sum_{\sigma \in X} \mu_N(\sigma) + N^{nq} \left(e^{-\delta N} + e^{-c_{10}(n)N} + |\mathcal{M}^n| e^{-c_{11}(n)N} \right) \mu_N[A]$$

$$\leq \mu_N[\mathcal{V}^n(A)] + N^{nq} \left(e^{-\delta N} + e^{-c_{10}(n)N} + |\mathcal{M}^n| e^{-c_{11}(n)N} \right) \mu_N[A], \quad (3.35)$$

where $c_{10}(n) := \mathfrak{C} - \varepsilon(n)c_1$ and $c_{11}(n) := \min\{\delta, \mathfrak{C} - \delta\} - \varepsilon(n)c_1$, while, by employing (3.30) and (3.31), the corresponding lower bound is given through

$$\sum_{\sigma \in S_N} \mu_N(\sigma) h_{A,B}(\sigma) \geq \sum_{X \in \mathcal{V}^n(A)} \sum_{\sigma \in X} \mu_N(\sigma) - \sum_{X \in \mathcal{V}^n(A)} \sum_{\sigma \in X} \mu_N(\sigma) \left(1 - h_{A,B}(\sigma)\right)$$

$$\geq \mu_N[\mathcal{V}^n(A)] - N^{nq} \left(e^{-\delta N} + e^{-c_{10}(n)N} \right) \mu_N[A]. \quad (3.36)$$

Hence, there exists $c_9 > 0$ such that for N large enough the assertion (3.29) follows. □

Corollary 3.12. *For $A \in \mathcal{M}^n$ and $B = \mathcal{M}^n(A)$ assume that the non-degeneracy condition is satisfied. Further, suppose that $n \ll N$ is such that $\varepsilon(n)c_1 < \mathfrak{C}/4$ where c_1 is chosen in such a way that $e^{-c_2 N} < \mathbb{P}_{\mu_A}[\tau_B < \tau_A]$. Then the conclusions of Theorem 3.10 also hold. Moreover,*

$$\mathbb{E}_\sigma[\tau_B] = \mathbb{E}_{\nu_{A,B}}[\tau_B] \left(1 + \mathcal{O}(e^{-c_0 N})\right), \quad \forall \sigma \in A. \quad (3.37)$$

Proof. By inspecting the proof of Theorem 3.10, we see that in the case when $B = \mathcal{M}^n(A)$ it suffices to ensure that $\varepsilon(n)c_1 < \mathfrak{C}$. Further, recall that for any metastable sets $A \in \mathcal{M}^n$ there exists $a \in \Gamma^n$ such that $A = S^n[a]$. Thus, for every $A \in \mathcal{M}^n$ and $B \subset \mathcal{M}^n \setminus A$ the condition (2.5) follows from (3.1), while (2.6) is redundant. Thus, as soon as $\varepsilon(n)c_1 < \mathfrak{C}/4$, Theorem 2.12 implies the pointwise estimate. □

3.4. Asymptotic exponential distribution

The content of this section is to show how the pointwise estimates on the Laplace transform of metastable exit times, which we obtained via coupling methods, combined with the averaged renewal equation (1.102) can be used to prove the convergence of the normalized metastable exit times to an exponential distribution, in cases where Theorem 3.10 applies.

3.4. Asymptotic exponential distribution

Theorem 3.13. *For $A \in \mathcal{M}^n$ and $B \subset \mathcal{M}^n(A)$ assume that the non-degeneracy condition is satisfied. Further, assume that $n \ll N$ is such that $\varepsilon(n)c_1 < \min\{\delta, \mathfrak{C}/4\}$ where δ is specified in (3.28) and c_1 is chosen in such a way that $\mathrm{e}^{-c_2 N} < \mathbb{P}_{\mu_A}[\tau_B < \tau_A]$. Then, for all $t \in \mathbb{R}_+$,*

$$\mathbb{P}_\sigma\left[\tau_B / \mathbb{E}_\sigma[\tau_B] > t\right] \longrightarrow \mathrm{e}^{-t}, \quad \text{as} \quad N \to \infty \tag{3.38}$$

for all $\sigma \in A$.

In view of Proposition 2.13, the main step in the proof is to show the convergence of the Laplace transform $\mathbb{E}_{\rho_\lambda}[\mathrm{e}^{-\frac{\lambda}{T}\tau_B}]$ as N tends to ∞, where the probability measure ρ_λ is defined in (1.101) and $T \equiv \mathbb{E}_{\nu_{A,B}}[\tau_B]$. By Lemma 1.25, our starting point is

$$\mathbb{E}_{\rho_\lambda}\left[\mathrm{e}^{-\frac{\lambda}{T}\tau_B}\right] = \frac{\mathbb{E}_{\mu_A}\left[\mathrm{e}^{-\frac{\lambda}{T}\tau_B} \mathbf{1}_{\tau_B < \tau_A}\right]}{1 - \mathbb{E}_{\mu_A}\left[\mathrm{e}^{-\frac{\lambda}{T}\tau_A} \mathbf{1}_{\tau_A < \tau_B}\right]}. \tag{3.39}$$

As a result of the representation (3.39) and the continuity theorem for Laplace transforms, see [49, XIII.1 Theorem 2], Theorem 3.13 will follow from (2.63) and (2.57), respectively, once we have proven the following lemma.

Lemma 3.14. *Under the assumptions of Theorem 3.13, for any $\lambda \geq 0$*

$$\lim_{N \to \infty} \frac{\mathbb{E}_{\mu_A}\left[\mathrm{e}^{-\frac{\lambda}{T}\tau_B} \mathbf{1}_{\tau_B < \tau_A}\right]}{1 - \mathbb{E}_{\mu_A}\left[\mathrm{e}^{-\frac{\lambda}{T}\tau_A} \mathbf{1}_{\tau_A < \tau_B}\right]} = \frac{1}{1+\lambda}. \tag{3.40}$$

Proof. The proof of this lemma comprises the following three steps.

STEP 1. First of all, we will prove a crucial bound to which we refer to as *Uphill Lemma*.

Lemma 3.15. *[5, Lemma 4.3] Under the assumptions of Theorem 3.13, there exists $c_{12} > 0$ and $N_0 \in \mathbb{N}$ such that*

$$\mathbb{E}_{\mu_A}\left[\tau_B \mathbf{1}_{\tau_B < \tau_A}\right] \leq \mathrm{e}^{-c_{12} N} \mathbb{E}_{\mu_A}\left[\tau_A \mathbf{1}_{\tau_A < \tau_B}\right], \quad \forall N \geq N_0. \tag{3.41}$$

Proof. To start with, we use the fact that

$$\mathbb{E}_{\mu_A}\left[\tau_{A \cup B}\right] = \mathbb{E}_{\mu_A}\left[\tau_A \mathbf{1}_{\tau_A < \tau_B}\right] + \mathbb{E}_{\mu_A}\left[\tau_B \mathbf{1}_{\tau_B < \tau_A}\right]. \tag{3.42}$$

Notice that, for $\sigma \in A$,

$$\mathbb{E}_\sigma\left[\tau_A \mathbf{1}_{\tau_A < \tau_B}\right] = \mathbb{P}_\sigma[\tau_A < \tau_B] + (L w_{A,B})(\sigma). \tag{3.43}$$

An application of the second Green's identity to the functions $w_{A,B}$ and $h_{A,B}$ reveals that

$$\sum_{\sigma \in A} \mu_N(\sigma) (L w_{A,B})(\sigma) = \sum_{\sigma \notin A \cup B} \mu_N(\sigma) (h_{A,B}(\sigma))^2. \tag{3.44}$$

Hence,
$$\mathbb{E}_{\mu_A}[\tau_A \mathbb{1}_{\tau_A<\tau_B}] = \mathbb{P}_{\mu_A}[\tau_A < \tau_B] + \frac{1}{\mu_N[A]} \sum_{\sigma \notin A \cup B} \mu_N(\sigma) \left(h_{A,B}(\sigma)\right)^2. \quad (3.45)$$

On the other hand, by an analog procedure, one shows that
$$\mathbb{E}_{\mu_A}[\tau_{A \cup B}] = 1 + \frac{1}{\mu_N[A]} \sum_{\sigma \in A} \mu_N(\sigma) \left(Lw_{A \cup B}\right)(\sigma)$$
$$= 1 + \frac{1}{\mu_N[A]} \sum_{\sigma \notin A \cup B} \mu_N(\sigma) h_{A,B}(\sigma). \quad (3.46)$$

Thus, taking into account (3.42) we can conclude that
$$\mathbb{E}_{\mu_A}[\tau_B \mathbb{1}_{\tau_B<\tau_A}] = \frac{\mathrm{cap}(A,B)}{\mu_N[A]} + \frac{1}{\mu_N[A]} \sum_{\sigma \notin A \cup B} \mu_N(\sigma) h_{A,B}(\sigma) h_{B,A}(\sigma). \quad (3.47)$$

Due to (3.1), the first term on the right-hand side is bounded from above by $\mathrm{e}^{-\mathfrak{e}N}$. Further, from Lemma 3.8 together with the non-degeneracy condition we get that the second term is as well exponentially small compared to $\mu_N[A]$. Moreover, since $\mathbb{E}_{\mu_A}[\tau_{A \cup B}] \geq 1$ we conclude that $\mathbb{E}_{\mu_A}[\tau_A \mathbb{1}_{\tau_A<\tau_B}]$ is exponentially close to one. Hence, choosing $c_{12} > 0$ and $N_0 \in \mathbb{N}$ appropriately, we deduce (3.41). □

Corollary 3.16. *With the notations introduced above, it holds that*
$$\mathbb{E}_{\nu_{A,B}}[\tau_B] = \frac{\mathbb{E}_{\mu_A}[\tau_A \mathbb{1}_{\tau_A<\tau_B}]}{\mathbb{P}_{\mu_A}[\tau_B < \tau_A]} \left(1 + \mathrm{e}^{-c_{12}N}\right), \quad \forall N \geq N_0. \quad (3.48)$$

Proof. As an immediate consequence of Lemma 1.14 combined with (3.41), we obtain (3.48). □

STEP 2. Using that, $1 - x \leq \mathrm{e}^{-x} \leq 1$ for $x \geq 0$, it follows that the numerator in the renewal equation (3.39) can be written as
$$\mathbb{P}_{\mu_A}[\tau_B < \tau_A] - \frac{\lambda}{T} \mathbb{E}_{\mu_A}[\tau_B \mathbb{1}_{\tau_B<\tau_A}] \leq \mathbb{E}_{\mu_A}\!\left[\mathrm{e}^{-\frac{\lambda}{T}\tau_B} \mathbb{1}_{\tau_B<\tau_A}\right] \leq \mathbb{P}_{\mu_A}[\tau_B < \tau_A]. \quad (3.49)$$

In view of the Uphill Lemma and (3.48), we get
$$\mathbb{P}_{\mu_A}[\tau_B < \tau_A]\left(1 - \lambda\mathrm{e}^{-c_{11}N}\right) \leq \mathbb{E}_{\mu_A}\!\left[\mathrm{e}^{-\frac{\lambda}{T}\tau_B} \mathbb{1}_{\tau_B<\tau_A}\right] \leq \mathbb{P}_{\mu_A}[\tau_B < \tau_A]. \quad (3.50)$$

Let us now turn to the denominator in (3.39). As a consequence of the second Green's identity applied to the function $h_{A,B}$ and $h^\lambda_{A,B}$, we may rewrite it as
$$1 - \mathbb{E}_{\mu_A}\!\left[\mathrm{e}^{-\frac{\lambda}{T}\tau_A} \mathbb{1}_{\tau_A<\tau_B}\right]$$
$$= \mathbb{P}_{\mu_A}[\tau_B < \tau_A]\,\mathrm{e}^{-\frac{\lambda}{T}} + \frac{1 - \mathrm{e}^{-\frac{\lambda}{T}}}{\mu_N[A]} \sum_{\sigma \notin B} \mu_N(\sigma) h_{A,B}(\sigma) h^\lambda_{A,B}(\sigma). \quad (3.51)$$

3.4. Asymptotic exponential distribution

Taking into account the representation (1.49) and using the fact that $h_{A,B}^\lambda(\sigma) \leq h_{A,B}(\sigma) \leq 1$ for all $\sigma \in S_N$ we get

$$1 - \mathbb{E}_{\mu_A}\left[e^{-\frac{\lambda}{T}\tau_A}\mathbb{1}_{\tau_A < \tau_B}\right] \leq \mathbb{P}_{\mu_A}[\tau_B < \tau_A](1+\lambda). \tag{3.52}$$

Hence, our remaining task is to derive a lower bound that is exponentially close to the upper bound. In order to do so, let us restrict the summation on the right-hand side of (3.51) to the subset $A_\delta \subset S^n$ that is given by the union of all sets in $\mathcal{U}_\delta^n(A)$. Moreover, δ it given by the non-degeneracy condition. This yields

$$1 - \mathbb{E}_{\mu_A}\left[e^{-\frac{\lambda}{T}\tau_A}\mathbb{1}_{\tau_A < \tau_B}\right]$$

$$\geq \mathbb{P}_{\mu_A}[\tau_B < \tau_A]\,e^{-\frac{\lambda}{T}}\left(1 + \frac{\lambda}{T\,\mathrm{cap}(A,B)}\sum_{\sigma \in A_\delta}\mu_N(\sigma)\,h_{A,B}(\sigma)\,h_{A,B}^\lambda(\sigma)\right) \tag{3.53}$$

where we used that $e^x - 1 \geq x$ for all $x \geq 0$.

STEP 3. We can conclude the proof of the Lemma 3.14, ones we have proven the following estimate.

Lemma 3.17. *Under the assumptions of Theorem 3.13 there exists $c_{13} > 0$ and $N_0 \in \mathbb{N}$ such that*

$$\sum_{\sigma \in A_\delta}\mu_N(\sigma)\,h_{A,B}(\sigma)\,h_{A,B}^\lambda(\sigma) \geq \sum_{\sigma \notin B}\mu_N(\sigma)\,h_{A,B}(\sigma)\left(1 - \lambda e^{-c_{13}N}\right), \qquad \forall N \geq N_0. \tag{3.54}$$

Proof. The idea behind the proof is to show that $h_{A,B}^\lambda$ and $h_{A,B}$ are almost of the same size on the subset $A_\delta \subset S_N$. To start with, we rewrite (3.54) as

$$\sum_{\sigma \in A_\delta}\mu_N(\sigma)\,h_{A,B}(\sigma)\,h_{A,B}^\lambda(\sigma)$$

$$\geq \sum_{\sigma \in A_\delta}\mu_N(\sigma)\,h_{A,B}(\sigma)^2 - \sum_{\sigma \in A_\delta}\mu_N(\sigma)\left(h_{A,B}(\sigma) - h_{A,B}^\lambda(\sigma)\right). \tag{3.55}$$

In view of Lemma 3.8 and 3.11, we immediately obtain, for N large enough, that

$$\sum_{\sigma \in A_\delta}\mu_N(\sigma)\,h_{A,B}(\sigma)^2$$

$$\geq \sum_{\sigma \notin B}\mu_N(\sigma)\,h_{A,B}(\sigma)\left(1 - N^{nq}\left(e^{-\delta N} + 2e^{-c_{10}(n)N} + |\mathcal{M}^n|e^{-c_{11}(n)N}\right)\right). \tag{3.56}$$

Let us point out that $c_{10}(n), c_{11}(n) > 0$ provided that n is chosen large enough such that $\varepsilon(n)c_1 < \min\{\delta, \mathfrak{C} - \delta\}$. Exploiting the probabilistic interpretation of $h_{A,B}$ and

$h_{A,B}^\lambda$, gives

$$h_{A,B}(\sigma) - h_{A,B}^\lambda(\sigma) = \mathbb{E}_\sigma\left[\left(1 - e^{-\frac{\lambda}{T}\tau_A}\right)\mathbb{1}_{\tau_A < \tau_B}\right] \leq \frac{\lambda}{T}\mathbb{E}_\sigma\left[\tau_A \mathbb{1}_{\tau_A < \tau_B}\right] \quad (3.57)$$

for all $\sigma \in A_\delta \setminus A$, whereas the difference between $h_{A,B}$ and $h_{A,B}^\lambda$ vanishes on A. Hence,

$$\sum_{\sigma \in A_\delta} \mu_N(\sigma)\left(h_{A,B}(\sigma) - h_{A,B}^\lambda(\sigma)\right) \leq \frac{\lambda}{T} \sum_{X \in \mathcal{U}_\delta^n(A) \setminus A} \sum_{\sigma \in X} \mu_N(\sigma) w_{A,B}(\sigma). \quad (3.58)$$

On the other hand, let us remark that for any $X \subset \mathcal{S}_N$, an application of the second Green's identity to the functions $w_{A,B}$ and $h_{X, A \cup B}$ yields

$$-\sum_{\sigma \in X} \mu_N(\sigma)(Lh_{X, A \cup B})(\sigma) w_{A,B}(\sigma) = \sum_{\sigma \notin A \cup B} \mu_N(\sigma) h_{X, A \cup B}(\sigma) h_{A,B}(\sigma). \quad (3.59)$$

Hence, by dividing both sides by $\text{cap}(X, A \cup B)$, we obtain

$$\mathbb{E}_{\nu_{X, A \cup B}}\left[\tau_A \mathbb{1}_{\tau_A < \tau_B}\right] = \frac{1}{\text{cap}(X, A \cup B)} \sum_{\sigma \notin A \cup B} \mu_N(\sigma) h_{X, A \cup B}(\sigma) h_{A,B}(\sigma), \quad (3.60)$$

where $\nu_{X, A \cup B}$ is the last exit biased distribution on X. In view of (2.27), we have for any $X \in \mathcal{U}_\delta^n(A) \setminus A$ and all $\sigma, \eta \in X$, that

$$\mathbb{P}_\sigma\left[\tau_{A \cup B} < \tau_X\right] \geq e^{-\varepsilon(n)c_1 N}\left(\mathbb{P}_\eta\left[\tau_{A \cup B} < \tau_X\right] - e^{-c_2 N}\right). \quad (3.61)$$

By multiplying both sides with $\mu_N(\eta)$ and summing over all $\eta \in X$, we obtain the following pointwise estimate

$$\mathbb{P}_\sigma\left[\tau_{A \cup B} < \tau_X\right] \geq e^{-\varepsilon(n)c_1 N} \frac{\text{cap}(A \cup B, X)}{\mu_N[X]}\left(1 - 2e^{-\mathfrak{C}N}\right). \quad (3.62)$$

Here, we used that $\mathbb{P}_{\mu_A}\left[\tau_B < \tau_A\right] \leq 2e^{-\mathfrak{C}N}\mathbb{P}_{\mu_X}\left[\tau_{A \cup B} < \tau_X\right]$ for all $X \in \mathcal{U}_\delta^n(A) \setminus A$ by (3.1), and that c_2 is chosen in such a way that $e^{-c_2 N} < \mathbb{P}_{\mu_A}\left[\tau_B < \tau_A\right]$. Hence, by combining the pointwise estimate with (3.60) yields

$$\sum_{\sigma \in X} \mu_N(\sigma) w_{A,B}(\sigma)$$

$$\leq e^{\varepsilon(n)c_1 N}\left(1 - 2e^{-\mathfrak{C}N}\right)^{-1} \frac{\mu_N[X]}{\text{cap}(X, A \cup B)} \sum_{\sigma \in \mathcal{S}_N} \mu_N(\sigma) h_{A,B}(\sigma)$$

$$\leq \frac{2e^{-N(\mathfrak{C}-\varepsilon(n)c_1)}}{\mathbb{P}_{\mu_A}\left[\tau_B < \tau_A\right]}\left(1 - 2e^{-\mathfrak{C}N}\right)^{-1} \sum_{\sigma \in \mathcal{S}_N} \mu_N(\sigma) h_{A,B}(\sigma). \quad (3.63)$$

Finally, by plugging (3.63) into (3.58), we obtain

$$\sum_{\sigma \in A_\delta} \mu_N(\sigma)\left(h_{A,B}(\sigma) - h_{A,B}^\lambda(\sigma)\right) \leq \lambda 4 N^{nq} e^{-N(\mathfrak{C}-\varepsilon(n)c_1)} \sum_{\sigma \notin B} \mu_N(\sigma) h_{A,B}(\sigma)$$

for N large enough. Thus, by combining the estimates above concludes the proof. $\square$

3.4. Asymptotic exponential distribution

From Lemma 3.17, it follows that the denominator of (3.39) can be bounded by

$$\left(1 - \lambda e^{-\mathfrak{C}N}\right)\left(1 + \lambda\left(1 - \lambda e^{-c_{13}N}\right)\right) \leq \frac{1 - \mathbb{E}_{\mu_A}\left[e^{-\frac{\lambda}{T}\tau_A}\mathbb{1}_{\tau_A < \tau_B}\right]}{\mathbb{P}_{\mu_A}\left[\tau_B < \tau_A\right]} \leq 1 + \lambda \tag{3.64}$$

provided that N is chosen large enough. Together with (3.50) the assertion of Lemma 3.14 follows. □

Chapter 4

Metastability and Spectral Theory

In the present chapter we study the relation between the low-lying part of the spectrum of $-L_N$ and the mean transition times associated to certain metastable sets in $\mathcal{M}^n$. The main objective is to show that, under some non-degeneracy condition, to each metastable set corresponds a simple eigenvalue of $-L_N$ which is equal to the inverse of the mean exit time from this set up to small errors. Let us point out that the methods we are going to use in the sequel are mainly analytic.

The remainder of this chapter is organized as follows. In Section 4.1 we review first the main ideas and methods that were used in the strong recurrent setting [16]. In particular, we explain some of the challenges, we are faced with, and sketch a strategy to solve them. A first observation is a relation between the eigenvalues of the generator $-L_N$ and the eigenvalues of the generator $-G^n$ associated to the coarse-grained Markov chain. By employing this observation, in Section 4.2 we prove a rough localization of the small eigenvalues. In the remaining Section 4.3 we focus on improving the previously obtained estimates. Using *a posteriori* error estimates based on general results due to Kato [74] and Temple [109] combined with properties of metastable sets and pointwise estimates on harmonic functions we derive upper and lower bounds on the small eigenvalues that coincides in the limit when N tends to infinity.

Contributions in this chapter. The main contribution of the author is the development of a strategy that allows to reduce the original eigenvalue problem on $\mathcal{S}_N$ to the strong recurrent setting on the coarse-grained level. A crucial step of the presented method is to establish first a rough localization of the eigenvalues. Under the assumption that there are at most as many small eigenvalues as metastable sets, these estimates can be used in a second step to compute precisely the small eigenvalues. Let us point out that this approach is not limited to the particular setting we have chosen. Beyond that, the chapter contains the following novel pieces:

- a lower bound on the eigenvalues of the generator $-L_N$ in terms of the eigenvalues of the generator $-G^n$, see Proposition 4.5;
- an upper bound on the residuum exploiting pointwise estimates on harmonic functions, see Lemma 4.14;
- an analysis of the Rayleigh-Ritz values that appear in the improved upper bound, see 4.13.

4.1. Introduction

Investigations in the connection between the dynamical behavior of Markov processes and the existence of small eigenvalues of the corresponding generators dates back at least to the work of Wentzell [112] and Feidlin and Wentzell [51]. In the study of diffusions, they identified $\lim_{\varepsilon \downarrow 0} \varepsilon^{-1} \ln \lambda(\varepsilon)$ using large deviation methods. Based on variational principles, these estimates could be improved up to a multiplicative error first by Holley, Kusuoka and Stroock [65] for principal eigenvalues and later for the full set of exponentially small eigenvalues by Miclo [86] and Mathieu [84].

By using potential-theoretic ideas that were already suggested in an early work by Wentzell [113], sharp estimates on the small eigenvalues were established by Bovier, Gayrard, Klein and Eckhoff for reversible diffusion processes [18, 45] and for Markov chains on discrete state spaces in the reversible setting [16]. In the context of reversible Markov chains on discrete state spaces, their starting point was the definition of a *set of metastable points*. Under some non-degeneracy conditions they showed that to each metastable point corresponds a simple eigenvalue of the discrete generator associated to the Markov chain. Moreover, each such eigenvalue is equal to the inverse of the mean exit time from the corresponding metastable point up to negligible error terms.

In the previous chapter we have seen, that e.g. in the context of stochastic spin systems at finite temperature, the probability to hit a particular configuration in the state space and the escape probability from a metastable point are of the same order. For this reason, a Markov chain in such a setting can only be ρ-metastable in the sense of [11, Definition 4.1], if the transition probabilities are exponentially small. For instance, this is the case if we consider a spin system in the low temperature regime. In contrast to that, mean-field systems can be studied as well in the finite temperature regime. The reason behind this fact is that mean-field systems have the property that there exists a macroscopic variable such that the stochastic process on the coarse-grained space induced by this map is still Markovian. Thus, the set of metastable points can be defined on this lower dimensional space. A further advantage of mean-field systems is that the

4.1. Introduction

spectrum of the discrete generator, $-G_N$, associated to the induced Markov chain on the lower-dimensional space is contained in the spectrum of $-L_N$.

As we already mentioned in the preface, our interest in the relation between the metastable behavior and the small eigenvalues of the generator associated to a Markov chain arises from the study of disordered mean-field spin systems at *finite temperature*. In such a situation, we *cannot* exactly reduce the model to a low-dimensional one via lumping techniques.

Before discussing the difficulties, we are faced with when starting from the definition of a set of metastable sets, let us briefly sketch the main ideas of the approach, originally presented in [16], to investigate the small eigenvalues for strongly recurrent, reversible Markov chains on a finite state spaces. For an in-deep presentation of the connection between metastability and small eigenvalues from a potential-theoretic point of view we refer to the lecture notes [9, 11].

Small eigenvalues in the strong recurrent setting: A review of the methods. Let us consider the setting that was introduced in Chapter 1, i.e. let $\{\sigma(t)\}$ be an irreducible Markov chain on a finite state space, $\mathcal{S}_N$, that is reversible with respect to a unique invariant measure μ_N, and assume additionally that $\{\sigma(t)\}$ is ρ-metastable in the sense of [11, Definition 4.1]. This means that, for $0 < \rho \ll 1$, there exists a set of *metastable points*(!), $\mathcal{M}_N \equiv \{\sigma^1, \ldots, \sigma^K\} \subset \mathcal{S}_N$, such that

$$\frac{\max_{\eta \in \mathcal{M}_N} \mathbb{P}_\eta[\tau_{\mathcal{M}_N \setminus \{\eta\}} < \tau_\eta]}{\min_{\eta \in \mathcal{S}_N \setminus \mathcal{M}_N} \mathbb{P}_\eta[\tau_{\mathcal{M}_N} < \tau_\eta]} \leq \rho.$$

It is well known that reversibility implies that the generator, L_N, is self-adjoint on the weighted space $L^2(\mathcal{S}_N, \mu_N)$. Hence, the spectrum of L_N is real and we can order the eigenvalues, λ_i, of $-L_N$ in increasing order. As an immediate consequence of the Theorem of Perron-Frobenius we have that

$$0 = \lambda_1 < \lambda_2 \leq \lambda_3 \leq \ldots \leq \lambda_{|\mathcal{S}_N|} \leq 2.$$

Our main objective is to compute the first K eigenvalues of $-L_N$. As a first step, let us consider the eigenvalue problem for $-L_N$ subject to zero boundary conditions on a given subset $D \subset \mathcal{S}_N$. In such a situation, the smallest eigenvalue, that we denote by $\lambda^D \equiv \lambda_1^D$, is called principal eigenvalue. Notice that, for every set D, the principal eigenvalue has the pleasant feature to give rise to an *a priori* estimate of the $|D| + 1$ smallest eigenvalue of $-L_N$. Our starting point is the Theorem of Courant-Fischer. It implies that

$$\lambda_i \leq \lambda_i^D \leq \lambda_{i+|D|}, \qquad \forall i = 1, \ldots, |\mathcal{S}_N| - |D|. \tag{4.1}$$

We use this observation in the following way: Let us consider an increasing sequence of sets $D_1 \subset D_2 \subset \ldots \subset D_K$ such that D_1 and for all $i \geq 1$ the difference $D_{i+1} \setminus D_i$ is each a singleton. Then (4.1) implies that

$$\lambda^{D_i} \leq \lambda_{i+1} \quad \forall\, i = 1, \ldots, K. \tag{4.2}$$

The Equation (4.2) is one of the main ingredients to characterize completely the K smallest eigenvalues of $-L_N$. Note that every increasing sequence of sets $\{D_i\}$ can be used to derive a lower bound on the corresponding eigenvalue. But, the quality of such lower bounds depends crucially on the chosen sets D_i. In the iteration procedure that is presented below the idea is to construct an increasing sequence of sets from the set of metastable points.

Since the eigenvalue $\lambda_{|\mathcal{M}_N|+1}$ is bounded from below by the principal eigenvalue of $\lambda^{\mathcal{M}_N}$, a next step in the analysis is to derive bounds on it. Another pleasant feature of principal eigenvalues is that they can be easily characterized via variational principles. Consider a set $D \subset \mathcal{S}_N$. While an upper bound on λ^D follows from the Rayleigh principle, that reads

$$\lambda^D = \inf_{\substack{f|_D = 0 \\ \|f\|_{\mu_N} = 1}} \langle f, -L_N f \rangle_{\mu_N}, \tag{4.3}$$

a lower bound can be obtained by using a variational formula due to Donsker and Varadhan [40]

$$\lambda^D = \inf_{\substack{f|_D = 0 \\ \|f\|_{\mu_N} = 1}} \sup_{g|_{D^c} > 0} \left\langle \frac{f^2}{g}, -L_N g \right\rangle_{\mu_N}. \tag{4.4}$$

Here, we introduced the notations $\|f\|_{2,\mu_N}^2 \equiv \langle f, f \rangle_{\mu_N} := \sum_{\sigma \in \mathcal{S}_N} \mu_N(\sigma) f(\sigma)^2$.

Let us now come back to the principal eigenvalue of $-L_N$. An upper bound on $\lambda^{\mathcal{M}_N}$ is obtained by plugging the normalized equilibrium potential $h_{\eta,\mathcal{M}_N}/\|h_{\eta,\mathcal{M}_N}\|_{2,\mu_N}^2$ for some $\eta \in \mathcal{S}_N \setminus \mathcal{M}_N$ in the Rayleigh principle and optimizing over all η. For a lower bound, recall that the function $w_{\mathcal{M}_N}$ equals $\mathbb{E}_\sigma[\tau_{\mathcal{M}_N}]$ on $\mathcal{M}_N^c$. In particular, for all $\sigma \notin \mathcal{M}_N$ it holds that $(L_N w_{\mathcal{M}_N})(\sigma) = -1$. Thus, by setting $g \equiv w_{\mathcal{M}_N}$ the variational principle of Donsker-Varadhan yields that

$$\lambda^{\mathcal{M}_N} \geq \left(\max_{\sigma \in \mathcal{S}_N \setminus \mathcal{M}_N} \mathbb{E}_\sigma[\tau_{\mathcal{M}_N}] \right)^{-1}. \tag{4.5}$$

This proves the following

Lemma 4.1. *Let $\lambda^{\mathcal{M}_N}$ denote the principal Dirichlet eigenvalue of $-L_N$ this respect to the set $\mathcal{M}_N$. Then*

$$\min_{\sigma \in \mathcal{S}_N \setminus \mathcal{M}_N} \frac{\mathrm{cap}(\sigma, \mathcal{M}_N)}{\|h_{\sigma, \mathcal{M}_N \setminus \{\sigma\}}\|_{1,\mu_N}} \leq \lambda^{\mathcal{M}_N} \leq \min_{\sigma \in \mathcal{S}_N \setminus \mathcal{M}_N} \frac{\mathrm{cap}(\sigma, \mathcal{M}_N)}{\|h_{\sigma, \mathcal{M}_N \setminus \{\sigma\}}\|_{2,\mu_N}^2}. \tag{4.6}$$

4.1. Introduction

Having establish a lower bound for the principal Dirichlet eigenvalue, we turn, in a second step, to a precise characterization of the eigenvalues of $-L_N$ that are strictly below $\lambda^{\mathcal{M}_N}$. Notice that there are at most K such eigenvalues. The strategy is to consider the solution of the Dirichlet problem

$$\begin{cases} -(L_N f^\lambda)(\sigma) - \lambda f^\lambda(\sigma) = 0, & \sigma \in \mathcal{S}_N \setminus \mathcal{M}_N, \\ f^\lambda(\sigma) = \phi(\sigma), & \sigma \in \mathcal{M}_N. \end{cases} \quad (4.7)$$

If $\lambda < \lambda^{\mathcal{M}_N}$ is an eigenvalue of $-L_N$ and if we choose ϕ^λ as the eigenfunction corresponding to this eigenvalue, then f^λ equals ϕ^λ on $\mathcal{S}_N$. In order to see this, notice that $f^\lambda(\sigma) - \phi^\lambda(\sigma) = 0$ for all $\sigma \in \mathcal{M}_N$. Hence, the difference $f^\lambda - \phi^\lambda$ solves a Dirichlet problem with respect to the operator $-L_N - \lambda$ subject to zero boundary conditions. Due to the fact that $\lambda < \lambda^{\mathcal{M}_N}$, this Dirichlet problem has a unique solution given by the zero function. This implies that f^λ is equal to the eigenfunction ϕ^λ everywhere.

Notice that by linearity of the Dirichlet problem and due to the properties of the λ-equilibrium potential that was defined in Chapter 1, we can represent f^λ as

$$f^\lambda(\sigma) = \sum_{\eta \in \mathcal{M}_N} c_\eta \, h^\lambda_{\eta, \mathcal{M}_N \setminus \eta}(\sigma), \qquad c_\eta \in \mathbb{R}. \quad (4.8)$$

If for a given $\lambda < \lambda^{\mathcal{M}_N}$ there exists a non-zero vector $\{c_\eta : \eta \in \mathcal{M}_N\}$ such that for all $\sigma \in \mathcal{S}_N$ $-(L_N f^\lambda)(\sigma) = \lambda f^\lambda(\sigma)$, then λ is an eigenvalue of $-L_N$. The existence of such a vector is equivalent to the fact that the determinant of the matrix $E^\lambda_{\mathcal{M}_N} \in \mathbb{R}^{K \times K}$ with elements

$$\left(E^\lambda_{\mathcal{M}_N}\right)_{\sigma,\eta} = -\left(L_N h^\lambda_{\eta, \mathcal{M}_N \setminus \eta}\right)(\sigma) - \lambda h^\lambda_{\eta, \mathcal{M}_N \setminus \eta}(\sigma), \qquad \sigma, \eta \in \mathcal{M}_N. \quad (4.9)$$

vanishes. The next step in the analysis is to study the matrix $E^\lambda_{\mathcal{M}_N}$. Although this is in general a quite difficult task, if we assuming additionally the non-degeneracy condition (4.10), we can compute precisely the largest eigenvalues below $\lambda^{\mathcal{M}_N}$. This is the statement of the following

Theorem 4.2 ([11, Theorem 6.8]). *Assume that there exists $\eta \in \mathcal{M}_N$ such that for $0 < \delta \ll 1$*

$$\delta^2 \frac{\mathrm{cap}(\eta, \mathcal{M}_N \setminus \eta)}{\|h_{\eta, \mathcal{M}_N \setminus \eta}\|^2_{2,\mu_N}} \geq \max_{\sigma \in \mathcal{M}_N \setminus \{\sigma\}} \frac{\mathrm{cap}(\sigma, \mathcal{M}_N \setminus \sigma)}{\|h_{\sigma, \mathcal{M}_N \setminus \sigma}\|^2_{2,\mu_N}}. \quad (4.10)$$

Then, the largest eigenvalue λ_η of $-L_N$ below $\lambda^{\mathcal{M}_N}$ is given by

$$\lambda_\eta = \frac{\mathrm{cap}(\eta, \mathcal{M}_N \setminus \eta)}{\|h_{\eta, \mathcal{M}_N \setminus \eta}\|^2_{2,\mu_N}} \left(1 + \mathcal{O}(\rho^2 + \delta^2)\right) = \frac{1}{\mathbb{E}_\eta[\tau_{\mathcal{M}_N \setminus \eta}]} \left(1 + \mathcal{O}(\rho^2 + \delta^2)\right). \quad (4.11)$$

Based on this theorem and (4.2), we are now in the position to characterize iteratively all eigenvalues of $-L_N$ below $\lambda^{\mathcal{M}_N}$. In order to do so, consider an increasing sequence of metastable sets $\mathcal{M}_1 \subset \mathcal{M}_2 \subset \ldots \subset \mathcal{M}_K \equiv \mathcal{M}_N$ such that $\mathcal{M}_1 = \eta^1$

and $\mathcal{M}_i \setminus \mathcal{M}_{i-1} = \eta^i$ for all $1 < i \leq K$. We assume that this increasing sequence can be constructed in such a way that for all $1 < i \leq K$, η^i and $\mathcal{M}_{i-1}$ satisfy the non-degeneracy condition (4.10), i.e.

$$\delta^2 \frac{\text{cap}(\eta^i, \mathcal{M}_{i-1})}{\|h_{\eta^i, \mathcal{M}_{i-1}}\|_{2,\mu_N}^2} \geq \max_{\sigma \in \mathcal{M}_i} \frac{\text{cap}(\sigma, \mathcal{M}_i \setminus \sigma)}{\|h_{\sigma, \mathcal{M}_i \setminus \sigma}\|_{2,\mu_N}^2}, \qquad \forall i = 2, \ldots, K. \qquad (4.12)$$

Then, $-L_N$ has exactly K eigenvalues below $\lambda^{\mathcal{M}_N}$ given by

$$\lambda_1 = 0, \qquad \lambda_i = \frac{\text{cap}(\eta^i, \mathcal{M}_{i-1})}{\|h_{\eta^i, \mathcal{M}_{i-1}}\|_{2,\mu_N}^2} \left(1 + \mathcal{O}(\rho^2 + \delta^2)\right), \qquad \forall i = 2, \ldots, K. \qquad (4.13)$$

whereas the normalized eigenfunction corresponding to λ_i reads

$$\phi^{\lambda_i}(\sigma) = \frac{h_{\eta^i, \mathcal{M}_{i-1}}(\sigma)}{\|h_{\eta^i, \mathcal{M}_{i-1}}\|_{2,\mu_N}^2} + \mathcal{O}(\delta + \rho) \sum_{j=1}^{i-1} \frac{h_{\eta^j, \mathcal{M}_{j-1}}(\sigma)}{\|h_{\eta^j, \mathcal{M}_{j-1}}\|_{2,\mu_N}^2}. \qquad (4.14)$$

Challenges and outline of the strategy. The starting point for our further investigations in the small eigenvalues of the generator, $-L_N$, is the set of metastable sets, $\mathcal{M}^n$, that was introduced in Chapter 3. In the sequel, we discuss the difficulties that arise in this setting. In particular, we outline a strategy to overcome some of these problems.

Suppose that $\mathcal{M}^n$ contains K metastable sets. Let us stress the fact that, typically, the number of configurations in such metastable sets, that we denote by $|\mathcal{M}^n|$, is exponentially large in N. By (4.2), the principal eigenvalue, $\lambda^{\mathcal{M}^n}$, of $-L_N$ subject to zero boundary conditions on the set of configurations in $\mathcal{M}^n$ is a lower bound for $\lambda_{|\mathcal{M}^n|+1}$. Although we expect that there are only K eigenvalues below $\lambda^{\mathcal{M}^n}$ that are separated by a gap from the remaining eigenvalues in the spectrum, we do not have a priori such a knowledge at hand. This is the first crucial difference compared to the strong recurrent setting. Since $K \ll |\mathcal{M}^n|$, it might be that there are exponentially many small eigenvalues below $\lambda^{\mathcal{M}^n}$. Hence, even if we determine for each metastable set $M \in \mathcal{M}^n$ an eigenvalue that is below $\lambda^{\mathcal{M}^n}$, we cannot characterize fully the low-lying spectrum.

Open Problem 4.3. *Prove/disprove that there exists $0 < \delta < \mathfrak{C}$ such that $e^{-\delta N} \lambda^{\mathcal{M}^n} < \lambda_{K+1}$.*

Let us assume the existence of such a δ. In the sequel, we address the question of characterizing precisely the small eigenvalues. In order to derive necessary conditions on the low-lying spectrum of $-L_N$, a possible strategy is to relate the small eigenvalues to a matrix that is to leading order equal to the so-called capacitance matrix. The former is constructed as follows. Fix an eigenvalue $\lambda < e^{-\delta N} \lambda^{\mathcal{M}^n}$ and denote by ϕ^λ the corresponding eigenfunction. Then, for $X, Y \in \mathcal{M}^n$, the entries of the matrix

4.1. Introduction

$E(\lambda, \phi^\lambda) \in \mathbb{R}^{K \times K}$ are given by

$$\left(E(\lambda, \phi^\lambda)\right)_{X,Y} = -\sum_{\sigma \in X} \mu_N(\sigma) \phi^\lambda(\sigma) \left((L_N h^\lambda_{\eta, \mathcal{M}_N \setminus \eta})(\sigma) + \lambda h^\lambda_{\eta, \mathcal{M}_N \setminus \eta}(\sigma)\right). \tag{4.15}$$

In contrast to (4.9), in the entries of the matrix $E(\lambda, \phi^\lambda)$ appear beside the eigenvalue λ, that we want to compute, the eigenfunction ϕ^λ which we do not know a priori.

Let us point out that this approach was used in the context of reversible diffusions [18, 45]. The reason for its success in that context is based on the following observation. A regularity analysis of the eigenfunction shows, that within balls, $B_\varepsilon(x)$, of radius ε around the metastable point x, ϕ^λ is either almost constant and, in particular, does not change its sign or its modulus is close to zero. Let us stress the fact that a suitable regularity theory for eigenfunctions of spin systems is still missing. In particular, to establish that ϕ^λ does not change its sign within a metastable set X is rather challenging. For this reason, we propose a different strategy.

As a first step, we investigate the relation between the spectrum of $-L_N$ and the spectrum of the generator $-G^n$ associated to a canonical Markov chain $\{\varrho^n(t)\}$ on the lower dimensional state space Γ^n. Our goal is to find for each eigenvalue $\eta \in \mathrm{spec}(-G^n)$ an interval that contains at least one eigenvalue $\lambda \in \mathrm{spec}(-L_N)$. For this purpose, we construct an artificial Markov chain $\{\bar\sigma(t)\}$ on $\mathcal{S}_N$ that is exactly lumpable, i.e. $\varrho^n(\bar\sigma(t))$ is a Markov chain on Γ^n that coincides in law with the Markov chain $\{\varrho^n(t)\}$. In particular, the spectral properties of the generator associated to $\{\bar\sigma(t)\}$ are used to establish lower bounds on eigenvalues $\lambda \in \mathrm{spec}(-L_N)$.

In a second step we improve both the upper and the lower bounds for the low-lying eigenvalues. While the upper bound follows immediately by an analysis of certain Rayleigh-Ritz values, lower bounds, that coincide with the upper bounds asymptotically, are obtained by means of the following

Theorem 4.4 (Kato and Temple). *Let u be an approximate eigenvector of unit length and $\xi = \langle u, -L_N u \rangle_{\mu_N}$. Assume that there exists an interval (a, b) that contains ξ and exactly one eigenvalue λ of $-L_N$. Then*

$$-\frac{\|(L_N + \xi)u\|^2_{2,\mu_N}}{\xi - a} \leq \xi - \lambda \leq \frac{\|(L_N + \xi)u\|^2_{2,\mu_N}}{b - \xi}. \tag{4.16}$$

Proof. See [104, Theorem 3.8]. □

A key observation, see Lemma 4.14, is that for a suitable approximated eigenfunction, u, the residual $\|(L_N + \xi)u\| \leq e^{2\varepsilon(n)c_1 N} \xi$. Note that the exponential growing factor can be compensated by choosing n appropriately because ξ is exponentially small in N. Let us stress the fact that, under the assumption that $e^{\delta N} \lambda^{\mathcal{M}^n} \leq \lambda_{K+1}$ and a suitable non-degeneracy condition, the first K intervals are disjoint and contain exactly one eigenvalue of $-L_N$.

4.2. Rough localization of eigenvalues

The main objective in this section is to derive *a priori* estimates on the eigenvalues of $-L_N$. Recall that $\{\varrho^n(t)\}$ is a Markov chain on the coarse-grained space Γ^n which is reversible with respect to $\mathcal{Q}^n = \mu_N \circ (\varrho^n)^{-1}$. Its transition probabilities are given by

$$r^n(\boldsymbol{x}, \boldsymbol{y}) = \frac{1}{\mathcal{Q}^n(\boldsymbol{x})} \sum_{\sigma \in \mathcal{S}^n[\boldsymbol{x}]} \mu_N(\sigma) \sum_{\eta \in \mathcal{S}^n[\boldsymbol{y}]} p_N(\sigma, \eta)$$

Further, we associate to $\{\varrho^n(t)\}$ the generator, G^n, that acts on functions $g: \Gamma^n \to \mathbb{R}$ as

$$(G^n g)(\boldsymbol{x}) = \sum_{\boldsymbol{y} \in \Gamma^n} r^n(\boldsymbol{x}, \boldsymbol{y}) \left(g(\boldsymbol{y}) - g(\boldsymbol{x}) \right). \tag{4.17}$$

Let $\langle \cdot, \cdot \rangle_{\mathcal{Q}^n}$ stands for the scalar product that is defined by

$$\langle f, g \rangle_{\mathcal{Q}^n} := \sum_{\boldsymbol{x} \in \Gamma^n} \mathcal{Q}^n(\boldsymbol{x}) f(\boldsymbol{x}) g(\boldsymbol{x})$$

and we write $L^2(\Gamma^n, \mathcal{Q}^n)$ for the vector space $\mathbb{R}^{|\Gamma^n|}$ equipped with the scalar product $\langle \cdot, \cdot \rangle_{\mathcal{Q}^n}$. Since reversibility implies that the generator, G^n, is self-adjoint on $L^2(\Gamma^n, \mathcal{Q}^n)$, the spectrum of G^n is real. Let us order the eigenvalues increasingly. An immediate consequence of the Perron-Frobenius Theorem is that

$$0 = \eta_1 < \eta_2 \leq \eta_3 \leq \ldots \leq \lambda_{|\Gamma^n|} \leq 2 \tag{4.18}$$

Since we regard elements of $\mathbb{R}^{|\Gamma^n|}$ as functions from Γ^n to $\mathbb{R}$, we call eigenvectors of the matrix G^n eigenfunctions and vice versa.

Let us now consider the following two matrices $C = (c(\sigma, \boldsymbol{x})), D = (d(\sigma, \boldsymbol{x})) \in \mathbb{R}^{|\mathcal{S}_N| \times |\Gamma^n|}$ to which we refer to as *collection matrix* and *distribution matrix*. The elements of these matrices are defined by

$$c(\sigma, \boldsymbol{x}) := \mathbb{1}_{\sigma \in \mathcal{S}^n[\boldsymbol{x}]}, \qquad d(\sigma, \boldsymbol{x}) := \frac{\mu_N(\sigma)}{\mathcal{Q}^n(\boldsymbol{x})} \mathbb{1}_{\sigma \in \mathcal{S}^n[\boldsymbol{x}]}. \tag{4.19}$$

Taking advantage of these matrices, we have that $G^n = D^{\mathrm{T}} L_N C$. Moreover, for all functions $f: \mathcal{S}_N \to \mathbb{R}$ and $g: \Gamma^n \to \mathbb{R}$ it holds that

$$\langle g, D^{\mathrm{T}} f \rangle_{\mathcal{Q}^n} = \sum_{\boldsymbol{x} \in \Gamma^n} \mathcal{Q}^n(\boldsymbol{x}) g(\boldsymbol{x}) \sum_{\sigma \in \mathcal{S}_N} \frac{\mu_N(\sigma)}{\mathcal{Q}^n(\boldsymbol{x})} \mathbb{1}_{\sigma \in \mathcal{S}^n[\boldsymbol{x}]} f(\sigma)$$

$$= \sum_{\sigma \in \mathcal{S}_N} \mu_N(\sigma) f(\sigma) \sum_{\boldsymbol{x} \in \Gamma^n} \mathbb{1}_{\sigma \in \mathcal{S}^n[\boldsymbol{x}]} g(\boldsymbol{x}) = \langle Cg, f \rangle_{\mu_N}. \tag{4.20}$$

With the notations introduced above, we prove the following

4.2. Rough localization of eigenvalues

Proposition 4.5. *Let $\lambda_i \in \mathrm{spec}(-L_N)$ and $\eta_i \in \mathrm{spec}(-G^n)$ sorted in increasing order. Then,*

$$\lambda_1 = \eta_1 = 0, \qquad \lambda_i \leq \eta_i, \qquad \forall\, i = 2, \ldots, |\Gamma^n|. \tag{4.21}$$

Moreover, for every $i = 2, \ldots, |\Gamma^n|$ there exists $j_i \geq i$ with $j_i \neq j_{i'}$ for $i \neq i'$ such that

$$e^{-2\varepsilon(n)N} \eta_i \leq \lambda_{j_i} \leq e^{2\varepsilon(n)N} \eta_i. \tag{4.22}$$

Proof. Let us start with showing the upper bound in (4.21). By exploiting the Theorem of Courant and Fischer we immediately obtain that

$$\eta_i = \min_{\substack{V \subset \mathbb{R}^{|\Gamma^n|} \\ \dim V = i}} \max_{\substack{g \in V \\ g \neq 0}} \frac{\langle g, -G^n g\rangle_{Q^n}}{\langle g, g\rangle_{Q^n}} = \min_{\substack{V \subset \mathbb{R}^{|\Gamma^n|} \\ \dim V = i}} \max_{\substack{f \in CV \\ f \neq 0}} \frac{\langle f, -L_N f\rangle_{\mu_N}}{\langle f, f\rangle_{\mu_N}}$$

$$\geq \min_{\substack{W \subset \mathbb{R}^{|\mathcal{S}_N|} \\ \dim W = i}} \max_{\substack{f \in W \\ f \neq 0}} \frac{\langle f, -L_N f\rangle_{\mu_N}}{\langle f, f\rangle_{\mu_N}} = \lambda_i, \tag{4.23}$$

where we used in the second step that $\langle g, G^n g\rangle_{Q^n} = \langle Cg, L_N Cg\rangle_{\mu_N}$.

Next, we focus on proving the bounds described in (4.22). Let us stress the fact that in later applications we are particularly interested in the lower bound. To start with, let us sketch the strategy behind our proof. The first idea is to consider an additional irreducible Markov chain $\{\bar{\sigma}(t)\}$ on the state space $\mathcal{S}_N$ whose transition probabilities $\bar{p}_N$ are reversible with respect to a unique invariant measure $\bar{\mu}_N$. We denote its generator by $\bar{L}^n$. For a given coarsening level n, this Markov chain is constructed in such a way that the induced process $\{\varrho^n(\bar{\sigma}(t))\}$ is a Markov chain. As a consequence, we obtain that the eigenvalues of $-G^n$ are contained in the spectrum of the generator $-\bar{L}^n$. In a second step, we compare the eigenvalues of $-L_N$ and $-\bar{L}_N$, by exploiting the positivity of the terms in the Dirichlet form.

In order to define the Markov chain $\{\bar{\sigma}(t)\}$, we choose the transition probabilities $\bar{p}_N(\sigma, \eta)$ and the probability measure $\bar{\mu}_N(\sigma)$ for $\sigma, \eta \in \mathcal{S}_N$ in the following way. For $\boldsymbol{x}, \boldsymbol{y} \in \Gamma^n$ and $\sigma \in \mathcal{S}^n[\boldsymbol{x}], \eta \in \mathcal{S}^n[\boldsymbol{y}]$ set

$$\bar{p}_N(\sigma, \eta) = \frac{r^n(\boldsymbol{x}, \boldsymbol{y})}{|A_{\boldsymbol{x},\boldsymbol{y}}(\sigma)|} \mathbb{1}_{\bar{p}_N(\sigma, \eta) \neq 0} \quad \text{and} \quad \bar{\mu}_N(\sigma) = \frac{Q^n(\boldsymbol{x})}{|\mathcal{S}^n[\boldsymbol{x}]|}, \tag{4.24}$$

where $A_{\boldsymbol{x},\boldsymbol{y}}(\sigma)$ for $\sigma \in \mathcal{S}^n[\boldsymbol{x}]$ is a short hand notation for the set $\{\eta \in \mathcal{S}^n[\boldsymbol{y}] \mid d_H(\sigma, \eta) = 1\}$. As a consequence of the single site dynamics, the cardinality of $A_{\boldsymbol{x},\boldsymbol{y}}(\sigma)$ is constant on $\mathcal{S}^n[\boldsymbol{x}]$. In particular, it holds, for all $\boldsymbol{x}, \boldsymbol{y} \in \Gamma^n$, that

$$|\mathcal{S}^n[\boldsymbol{x}]| \, |A_{\boldsymbol{x},\boldsymbol{y}}(\sigma)| = |\mathcal{S}^n[\boldsymbol{y}]| \, |A_{\boldsymbol{y},\boldsymbol{x}}(\eta)| \qquad \forall \sigma \in \mathcal{S}^n[\boldsymbol{x}], \eta \in \mathcal{S}^n[\boldsymbol{y}]. \tag{4.25}$$

As an immediate consequence, the detailed balance condition is satisfied. Further notice that,

$$\sum_{\eta \in \mathcal{S}^n[\boldsymbol{y}]} \bar{p}_N(\sigma, \eta) = \sum_{\eta \in \mathcal{S}^n[\boldsymbol{y}]} \bar{p}_N(\sigma', \eta) = r^n(\boldsymbol{x}, \boldsymbol{y}), \quad \forall \, \sigma, \sigma' \in \mathcal{S}^n[\boldsymbol{x}]. \quad (4.26)$$

Since (4.26) holds true, employing the classical theorem of Burke and Rosenblatt [21], we can conclude that $\{\varrho^n(\bar{\sigma}(t))\}$ is a Markov chain.

Moreover, (4.26) implies that the subspace $\mathcal{C}$, spanned by the columns of the matrix C, is invariant under the linear map $\bar{L}_N$, i.e. $\bar{L}_N \mathcal{C} \subset \mathcal{C}$. Thus, $\mathrm{spec}(-G^n) \subset \mathrm{spec}(-\bar{L}_N)$. In order to see this, notice that $\bar{L}_N$ is $\mathcal{C}$-invariant if and only if there exists a matrix $M \in \mathbb{R}^{|\Gamma^n| \times |\Gamma^n|}$ such that $\bar{L}_N C = CM$. But, from this equation follows easily that Cg is an eigenfunction of $\bar{L}_N$ if g is an eigenfunction of M. Since, $G^n = D^T \bar{L}_N C = D^T C = M$ by construction, $G^n = M$. Notice that this relation between the eigenvalues was proven in [82, Lemma 12.8] by similar arguments. Although to each $\eta \in \mathrm{spec}(-G^n)$ there exists an unique $\lambda \in \mathrm{spec}(-\bar{L}_N)$, we do not know to which eigenvalue of $-L_N$ the eigenvalue $\eta_i \in \mathrm{spec}(-G^n)$ belongs to. In view of the Courant-Fischer Theorem, we have that for all $i = 2, \ldots, |\Gamma^n|$ there exists $j_i \geq i$ such that

$$\eta_1 = \bar{\lambda}_1 = 0, \quad \text{and} \quad \eta_i = \bar{\lambda}_{j_i}, \quad \forall \, i = 2, \ldots, |\Gamma^n|. \quad (4.27)$$

We are left with the task to establish a relation between the eigenvalues of $-L_N$ and $-\bar{L}_N$. Suppose that n is chosen large enough to ensure that $\varepsilon(n) \leq 3/5$. Then, the assumption (2.4) implies that for $\sigma \in \mathcal{S}^n[\boldsymbol{x}]$ and $\eta \in \mathcal{S}^n[\boldsymbol{y}]$ such that $p_N(\sigma, \eta) > 0$

$$e^{-\varepsilon(n)} \frac{p_N(\sigma, \eta)}{\bar{p}_N(\sigma, \eta)} \leq e^{\varepsilon(n)} \quad \text{and} \quad e^{-\varepsilon(n)(N-1)} \leq \frac{\mu_N(\sigma)}{\bar{\mu}_N(\sigma)} \leq e^{\varepsilon(n)(N-1)}. \quad (4.28)$$

Recall that the Dirichlet form, $\mathcal{E}(f)$, that is defined by $\mathcal{E}(f) = \langle f, -L_N f \rangle_{\mu_N}$, can be written as a positive quadratic form. Hence,

$$\langle f, -L_N f \rangle_{\mu_N} \geq e^{-\varepsilon(n)N} \frac{1}{2} \sum_{\sigma, \eta \in \mathcal{S}_N} \bar{\mu}_N(\sigma) \bar{p}_N(\sigma, \eta) (f(\sigma) - f(\eta))^2$$

$$= e^{-\varepsilon(n)N} \langle f, -\bar{L}_N f \rangle_{\bar{\mu}_N}, \quad (4.29)$$

whereas $\langle f, f \rangle_{\mu_N} \leq e^{\varepsilon(n)N} \langle f, f \rangle_{\bar{\mu}_N}$. A further application of the Courant-Fischer Theorem, yields

$$\lambda_{j_i} \geq e^{-2\varepsilon(n)N} \min_{\substack{W_{j_i} \subset \mathbb{R}^{|\mathcal{S}_N|} \\ \dim W_{j_i} = j_i}} \max_{\substack{f \in \mathcal{W}_i \\ f \neq 0}} \frac{\langle f, -\bar{L}_N f \rangle_{\bar{\mu}_N}}{\langle f, f \rangle_{\bar{\mu}_N}} = e^{-2\varepsilon(n)N} \bar{\lambda}_{j_i}. \quad (4.30)$$

Hence, by combining (4.27) and (4.30), the lower bound in (4.22) is immediate. Moreover, by interchanging the role of L_N and $\bar{L}_N$ in (4.29) we can conclude the proof. □

4.2. Rough localization of eigenvalues

Remark 4.6. Notice that the bounds $\lambda_i \leq \eta_i \leq \lambda_{|S_N|-|\Gamma^n|+i}$ for all $i = 2, \ldots, |\Gamma^n|$ have already been established in [60].

Corollary 4.7. *Suppose that there exists $K \in \{2, \ldots, |\Gamma^n|\}$ such that $\lambda_{K+1} > e^{2\varepsilon(n)N}\eta_K$. Then,*

$$e^{-2\varepsilon(n)N}\eta_i \leq \lambda_i \leq \eta_i, \qquad \forall\, i = 2, \ldots, K. \tag{4.31}$$

Proof. In view (4.22), every interval $[e^{-2\varepsilon(n)N}\eta_i, e^{2\varepsilon(n)N}\eta_i]$, for $i = 2, \ldots, |\Gamma^n|$, contains at least one eigenvalue of $-L_N$. By inspecting the proof of Proposition 4.5, we see that the assumption $\lambda_{K+1} > e^{2\varepsilon(n)N}\eta_K$ together with (4.27) implies that $\bar{\lambda}_K = \eta_K$. Since $j_i \geq i$, it follows that $\bar{\lambda}_i = \eta_i$ for all $i = 1, \ldots, K$ and (4.30) yields the assertion. □

Let us point out that it is a rather challenging task to prove that the number of eigenvalues below the principal Dirichlet eigenvalue $\lambda^{\mathcal{M}^n}$ coincide with the number of metastable sets, say K. To our best knowledge, an answer to this problem is so far missing. Notice that for the precise characterization of the small eigenvalues, that we present in the next section, it suffices to know a priori that there exists $\delta > 0$ such that $e^{-\delta N}\lambda^{\mathcal{M}^n} \leq \lambda_{K+1}$. Although differently formulated, we will assume in the sequel the existence of such a separation.

Lemma 4.8. *Suppose that the set of metastable sets, $\mathcal{M}^n$, contains $K \equiv K(N)$ elements, where $K(N)$ grows at most sub-exponentially. Assume there exists $0 < \mathfrak{c} < \mathfrak{C}$ such that*

$$\min_{X \in \mathcal{S}_N^n \setminus \mathcal{M}^n} \mathbb{P}_{\mu_X}\left[\tau_{\mathcal{M}^n} < \tau_X\right] \geq e^{-\mathfrak{c}N}, \tag{4.32}$$

where $\mathcal{S}_N^n = \{\mathcal{S}^n[x] \subset \mathcal{S}_N \mid x \in \Gamma^n\}$. Suppose that there exists $0 < \delta < \mathfrak{C} - \mathfrak{c}$ such that

$$\bar{\lambda}_{K+1} \geq e^{-\delta N}\bar{\lambda}^{\mathcal{M}^n}, \tag{4.33}$$

where $\bar{\lambda}^{\mathcal{M}^n}$ denotes the principal Dirichlet eigenvalue of the generator $-\bar{L}_N$ subject to zero boundary conditions on $\mathcal{M}^n$. If $n \ll N$ is chosen large enough to ensure that $2\varepsilon(n) < \mathfrak{C} - \mathfrak{c}$, then

$$\frac{1}{2KN^{n\cdot q}}e^{(\mathfrak{C}-\mathfrak{c}-\delta-2\varepsilon(n))N}\lambda_K \leq \lambda_{K+1}. \tag{4.34}$$

Proof. Recall that metastable sets are defined as preimages under ϱ^n of points $m \in M \subset \Gamma^n$.

As a first step we claim that $\bar{\lambda}^{\mathcal{M}^n} = \eta^M$, where η^M is the principal Dirichlet eigenvalue of $-G^n$. Let us denote by $\bar{L}_{\mathcal{M}^n}$ and G_M^n the operators subject to zero boundary conditions on $\mathcal{M}^n$ and M, respectively. We prove the claim above by contradiction. Assume that $\bar{\lambda}^{\mathcal{M}^n} < \eta^M$. In view of the discussion above, we know that $\mathrm{spec}(-G_M^n) \subset \mathrm{spec}(-\bar{L}_{\mathcal{M}^n})$. Hence, there exists $\bar{\lambda}_i^{\mathcal{M}^n} \in \mathrm{spec}(-\bar{L}_{\mathcal{M}^n})$ with $i > 1$ such that $\bar{\lambda}_i^{\mathcal{M}^n} = \eta^M$. By the Theorem of Perron-Frobenius we know that the eigenfunctions

ϕ^1 to $\bar{\lambda}^{\mathcal{M}^n}$ and g^1 to η^M are strictly positive. Since $\langle \phi^1, \phi^i \rangle_{\bar{\mu}_N} = 0$, the eigenfunction to $\bar{\lambda}_i^{\mathcal{M}^n}$ has to be somewhere negative. Since ϕ^i is constant for all $\sigma \in \mathcal{S}^n[x]$, see [82, Lemma 12.8] the eigenfunction g^1 is somewhere negative. Contradiction. Using a similar argument if η^M is assumed to be smaller than $\bar{\lambda}^{\mathcal{M}^n}$ proves the claim.

In view of (4.30) we have that

$$\lambda_{K+1} \geq e^{-2\varepsilon(n)N} \bar{\lambda}_{K+1} \geq e^{-(\delta+2\varepsilon(n))N} \eta^M. \tag{4.35}$$

Thus, the assertion follows once we have shown an upper bound on λ_K that is strictly smaller than a lower bound for $e^{-(\delta+2\varepsilon(n))N} \eta^M$. Combining (4.2) with (4.6) implies that

$$\eta^M \geq \min_{x \in \Gamma^n \setminus M} \frac{\mathrm{CAP}^n(x, M)}{\|h_{x,M}\|_{1,\mathcal{Q}^n}} \geq \frac{1}{N^{n \cdot q}} \left(\min_{x \in \Gamma^n \setminus M} \frac{\mathrm{CAP}^n(x, M)}{\mathcal{Q}^n(x)} \right)^2, \tag{4.36}$$

where we used in the second step [11, Lemma 4.8]. Furthermore, from the Dirichlet principle follows immediately that $\mathrm{CAP}^n(x, M) \geq \mathrm{cap}(X, \mathcal{M}^n)$ where $X = \mathcal{S}^n[x]$. Hence,

$$\eta^M \geq \frac{1}{N^{n \cdot q}} \left(\min_{X \in \mathcal{S}_N^n \setminus \mathcal{M}^n} \frac{\mathrm{cap}(X, \mathcal{M}^n)}{\mu_N[X]} \right)^2 \geq \frac{e^{-\varepsilon N}}{N^{n \cdot q}} \min_{X \in \mathcal{S}_N^n \setminus \mathcal{M}^n} \mathbb{P}_{\mu_X}[\tau_{\mathcal{M}^n} < \tau_X]. \tag{4.37}$$

Let us now derive an upper bound for λ_K that is strictly smaller than the right-hand side of (4.37). From (4.22) we already know that $\lambda_K < \eta_K$. Further, by (4.11), it holds that $\eta_K \leq \max_{x \in M} \mathrm{CAP}^n(x, M \setminus x)/\mathcal{Q}^n(x)$. However, instead of trying to bound from above the mesoscopic capacity $\mathrm{CAP}^n(x, M \setminus x)$ by the microscopic capacity $\mathrm{cap}(X, \mathcal{M}^n \setminus X)$, we present in the sequel an alternative proof of an upper bound for λ_K.

Let us consider the K-dimensional subspace $\mathcal{C}$ of $\mathbb{R}^{|\mathcal{S}^n|}$ that is spanned by normalized equilibrium potentials

$$\mathcal{C} = \mathrm{span}\left\{ h_{X, \mathcal{M}^n \setminus X} / \|h_{X, \mathcal{M}^n \setminus X}\|_{2, \mu_N}^2 \mid X \in \mathcal{M}^n \right\}.$$

By exploiting the Theorem of Courant and Fischer, we get

$$\lambda_K = \min_{\substack{W \subset \mathbb{R}^{|\mathcal{S}^n|} \\ \dim W = K}} \max_{\substack{f \in W \\ f \neq 0}} \frac{\langle f, -L_N f \rangle_{\mu_N}}{\langle f, f \rangle_{\mu_N}} \leq \max_{\substack{f \in \mathcal{C} \\ f \neq 0}} \frac{\langle f, -L_N f \rangle_{\mu_N}}{\langle f, f \rangle_{\mu_N}} = \max_{\substack{z \in \mathbb{R}^K \\ z \neq 0}} \frac{\langle z, Kz \rangle}{\langle z, Bz \rangle}, \tag{4.38}$$

where $K = (K_{X,Y}) \in \mathbb{R}^{K \times K}$ is the so-called *capacitance matrix* that reads

$$K_{X,Y} = \frac{\langle h_{X, \mathcal{M}^n \setminus X}, -L_N h_{Y, \mathcal{M}^n \setminus Y} \rangle_{\mu_N}}{\|h_{X, \mathcal{M}^n \setminus X}\|_{2, \mu_N} \|h_{Y, \mathcal{M}^n \setminus Y}\|_{2, \mu_N}} \qquad \forall X, Y \in \mathcal{M}^n \tag{4.39}$$

4.2. Rough localization of eigenvalues

while the matrix $B = (B_{X,Y}) \in \mathbb{R}^{K \times K}$ is given by

$$B_{X,Y} = \frac{\langle h_{X,\mathcal{M}^n \setminus X}, h_{Y,\mathcal{M}^n \setminus Y} \rangle_{\mu_N}}{\|h_{X,\mathcal{M}^n \setminus X}\|_{2,\mu_N} \|h_{Y,\mathcal{M}^n \setminus Y}\|_{2,\mu_N}} \qquad \forall\, X, Y \in \mathcal{M}^n. \tag{4.40}$$

Since both the matrix K and B are positive definite, we get that for all $z \in \mathbb{R}^K \setminus \{0\}$

$$\frac{\langle z, Kz \rangle}{\|z\|^2} \leq \lambda_{\max}(K) \leq \sum_{i=1}^{K} \lambda_i(K) = \operatorname{tr}(E) \leq K \max_{X \in \mathcal{M}^n} \mathbb{P}_{\mu_X}\left[\tau_{\mathcal{M}^n \setminus X} < \tau_X\right], \tag{4.41}$$

where we used in the last step that $\|h_{X,\mathcal{M}^n \setminus X}\|_{2,\mu_N}^2 \geq \mu_N[X]$. Furthermore, if we denote by $\lambda_{\min}(B)$ the smallest eigenvalue of B then $\langle z, Bz \rangle \geq \lambda_1(B) \|z\|^2$. By applying Gershgorin's Theorem, we obtain

$$\lambda_1(B) \geq \min_{X \in \mathcal{M}^n} \left(B_{X,X} - \sum_{Y \in \mathcal{M}^n \setminus X} |B_{X,Y}| \right)$$

$$\geq \min_{X \in \mathcal{M}^n} \left(1 - \sum_{Y \in \mathcal{M}^n \setminus X} \frac{\langle h_{X,\mathcal{M}^n \setminus X}, h_{Y,\mathcal{M}^n \setminus Y} \rangle_{\mu_N}}{\sqrt{\mu_N[X]\, \mu_N[Y]}} \right). \tag{4.42}$$

Suppose that $\mu_N[X] \leq \mu_N[Y]$. Otherwise, we exchange in the following argument X and Y. In view of Lemma 3.8, we have that $\langle h_{X,\mathcal{M}^n \setminus X}, h_{Y,\mathcal{M}^n \setminus Y} \rangle_{\mu_N}$ is exponentially small in N compared to $\mu_N[X]$. Since the number of metastable sets grows at most sub-exponentially, it holds that $\lambda_1(B) \geq 1 - o_N(1)$. Combining the estimates above and taking into account the properties of metastable sets, we get

$$\lambda_K \leq 2K e^{-\mathfrak{C}N} \min_{X \in \mathcal{S}_N^n \setminus \mathcal{M}^n} \mathbb{P}_{\mu_X}\left[\tau_{\mathcal{M}^n} < \tau_X\right] \tag{4.43}$$

Hence, by combining (4.35), (4.37) and (4.43), we conclude the proof. □

Remark 4.9. Suppose that we pick from each metastable set M_i a configuration $\eta^i \in M_i$. Then, by (4.2), we know that $\bar{\lambda}^M \leq \bar{\lambda}_{K+1}$ where $M = \{\eta^1, \ldots, \eta^K\}$. Hence, the problem concerning the number of eigenvalues of $-L_N$ below $\lambda^{\mathcal{M}^n}$ reduces to compare two principal Dirichlet eigenvalues, namely $\bar{\lambda}^M$ and $\bar{\lambda}^{\mathcal{M}^n}$. This has the following advantages:

1) Obviously, by Perron-Frobenius, the eigenfunction, ϕ_M, corresponding to $\bar{\lambda}^M$ is strictly positive.
2) There is a coupling of two versions of the Markov chain $\{\bar{\sigma}(t)\}$ starting in two different configurations $\sigma, \eta \in S^n[x]$ such that, for all later times t, $\varrho^n(\bar{\sigma}(t)) = \varrho^n(\bar{\eta}(t))$. Moreover, the Hamming distance between $\bar{\sigma}(t)$ and $\bar{\eta}(t)$ is non-increasing. Hence, the distribution of the coupling time decays exponentially. This observation may be helpful to control the oscillations of the eigenfunction, ϕ_M.

3) It suffices to show that there exist $0 < \delta < \mathfrak{C} - 2\varepsilon(n)$ such that $e^{-\delta N} \bar{\lambda}^{\mathcal{M}^n} \leq \bar{\lambda}^{M}$.

These observations may be helpful to prove (4.2) using only properties of metastable sets.

4.3. Characterization of small eigenvalues

In the present section we prove the main result of this chapter. Namely, starting from the rough localization of the spectrum of $-L_N$, we improve substantially both the upper and lower bounds for the low-lying part of the spectrum. Moreover, we establish a relation between the low-lying eigenvalues of $-L_N$ and the metastable exit times associated to certain metastable sets. In the sequel, we construct to each low-lying eigenvalue a pair of approximate eigenvalue/eigenfunction. Since the eigenvalues in the low-lying part of the spectrum are exponentially small in N, the strategy of our proof to derive *relative error bound* that allows to determine *a posteriori* the accuracy of an approximated eigenvalue.

Let us consider a set of metastable sets, $\mathcal{M}^n$, as defined in Definition 3.1. Suppose that $\mathcal{M}^n$ consists of K elements, i.e. $\mathcal{M}^n = \{M_1, \ldots, M_K\}$ with $M_i \cap M_j = \emptyset$ for $i \neq j$, where $K = K(N)$ may grow at most sub-exponential in N. In view of the discuss in the previous section, we assume in all what follows that there exists $\delta > 0$ such that $e^{-\delta N} \lambda^{\mathcal{M}^n} \leq \lambda_{K+1}$. Further, we take the following *minimal non-degeneracy* condition into account.

Assumption 4.10. *Let us assume that there exists $0 < \delta < \mathfrak{C}$ and an unique increasing sequence of sets of sets $\mathcal{M}_1^n \subset \mathcal{M}_2^n \subset \ldots \subset \mathcal{M}_K^n \equiv \mathcal{M}^n$ such that for all $i = 2, \ldots, K$, it holds that*

(i) $\mathcal{M}_{i-1}^n = \mathcal{M}_i^n \setminus X_i$, *where the set* $X_i \in \mathcal{M}_i^n$ *satisfies*

$$\max_{M \in \mathcal{M}_{i-1}^n} \mathbb{P}_{\mu_M}\left[\tau_{\mathcal{M}_i^n \setminus M} < \tau_M\right] < e^{-\delta N} \mathbb{P}_{\mu_{X_i}}\left[\tau_{\mathcal{M}_{i-1}^n} < \tau_{X_i}\right]; \quad (4.44)$$

(ii) *either*

$$\mu_N[M] < e^{-\delta N} \mu_N[X_i] \quad (4.45)$$

or

$$\mathbb{P}_{\mu_{X_i}}\left[\tau_{\mathcal{M}_{i-1}^n} < \tau_{X_i}\right] < e^{-\delta N} \mathbb{P}_{\mu_M}\left[\tau_{\mathcal{M}_{i-1}^n} < \tau_M\right] \quad (4.46)$$

for all $M \in \mathcal{M}^n \setminus \mathcal{M}_i^n$.

Remark 4.11. The asymptotic ultrametricity of metastable sets induces a tree structure, from which the properties of the long-time behavior of the process can be read off. Notice that, by taking into account the minimal non-degeneracy condition, we obtain a binary tree having the property that the length of the shorter sub-trees below a

4.3. Characterization of small eigenvalues

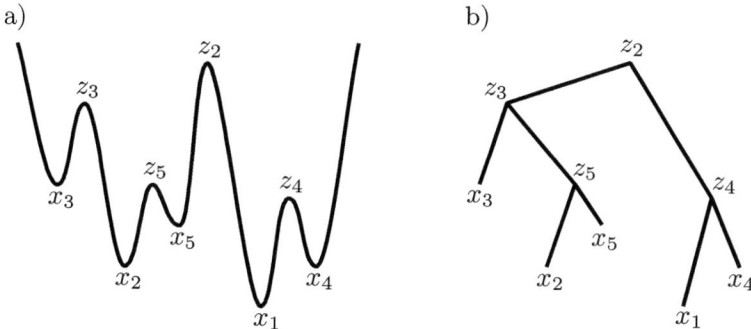

Figure 1. Illustration of a landscapes a) and the associated tree structure b). Since the landscape satisfies the minimal non-degeneracy condition, we can associate to each local minimum x_i a unique saddle point z_i. Moreover, the height differences between a local minimum and its saddle point are all distinct.

branching point are all distinct. The construction of this tree goes at follows: At the root split the tree into two sub-trees such that the metastable set X_1 is a leaf in the first and X_2 is a leaf in the second sub-tree. Within each of these two sub-trees we repeat this procedure, see Figure 4.3.

The assumption on λ_{K+1} together with the minimal non-degeneracy condition allows to characterize precisely the K smallest eigenvalues of $-L_N$.

Theorem 4.12. *Let us consider an increasing sequence $\mathcal{M}_1^n \subset \mathcal{M}_2^n \subset \ldots \subset \mathcal{M}_K^n \equiv \mathcal{M}^n$ that satisfies the minimal non-degeneracy condition specified in Assumption 4.10 and assume that there exists $0 < \mathfrak{c} < \mathfrak{C}$ such that*

$$\min_{X \in S_N^n \setminus \mathcal{M}^n} \mathbb{P}_{\mu_X}\left[\tau_{\mathcal{M}^n} < \tau_X\right] \geq e^{-\mathfrak{c} N}. \tag{4.47}$$

Suppose that there exists $0 < \delta < \mathfrak{C} - \mathfrak{c}$ such that $e^{-\delta N} \lambda^{\mathcal{M}^n} < \lambda_{K+1}$. If $n \ll N$ is large enough to ensure that $2\varepsilon(n)(c_1 + 1) < \min\{\mathfrak{C} - \mathfrak{c} - \delta, \delta\}$, where c_1 is chosen in such a way that $e^{-c_2 N} < \mathbb{P}_{\mu_{X_2}}\left[\tau_{X_1} < \tau_{X_2}\right]$. Then, there exists $c_{14} > 0$ such that

$$\lambda_1 = 0, \qquad \lambda_i = \frac{\operatorname{cap}(X_i, \mathcal{M}_{i-1}^n)}{\|h_{X_i, \mathcal{M}_{i-1}^n}\|_{2,\mu_N}^2}\left(1 - e^{-c_{14} N}\right), \qquad \forall i = 2, \ldots, K. \tag{4.48}$$

In particular, for all $i = 2, \ldots, K$,

$$\lambda_i = \frac{1}{\mathbb{E}_\sigma\left[\tau_{\mathcal{M}_{i-1}^n}\right]}\left(1 - e^{-c_{14} N}\right), \qquad \forall \sigma \in X_i. \tag{4.49}$$

Let us start with improving the upper bound given in Proposition 4.5.

Lemma 4.13. *Under the assumptions of Theorem 4.12, it holds that*

$$\lambda_1 = 0, \qquad \lambda_i \leq \frac{\operatorname{cap}(X_i, \mathcal{M}_{i-1}^n)}{\|h_{X_i, \mathcal{M}_{i-1}^n}\|_{2,\mu_N}^2} \left(1 + e^{-c_{14} N}\right), \qquad \forall i = 2, \ldots, K. \tag{4.50}$$

Proof. Note that the strategy of this proof is similar to the one given in Proposition 4.8. First, consider the vector $u^1, \ldots, u^K$ with $u^i \in \mathbb{R}^{|S_N|}$ that are given by

$$u^1 = 1, \qquad u^i = \frac{h_{X_i, \mathcal{M}_{i-1}^n}}{\|h_{X_i, \mathcal{M}_{i-1}^n}\|_{2,\mu_N}^2} \qquad \text{for } i = 2, \ldots, K. \tag{4.51}$$

Further, we denote by $\mathcal{C}_i$ the i-dimensional subspace $\mathcal{C}_i = \operatorname{span}\{u^1, \ldots, u^i\}$. Hence, for every $i = 1, \ldots, K$, we get

$$\lambda_i \leq \min_{\substack{W \subset \mathbb{R}^{|S^n|} \\ \dim W = i}} \max_{\substack{f \in W \\ f \neq 0}} \frac{\langle f, -L_N f\rangle_{\mu_N}}{\langle f, f\rangle_{\mu_N}} \leq \max_{\substack{f \in \mathcal{C}_i \\ f \neq 0}} \frac{\langle f, -L_N f\rangle_{\mu_N}}{\langle f, f\rangle_{\mu_N}} = \max_{\substack{z \in \mathbb{R}^i \\ z \neq 0}} \frac{\langle z, K_i z\rangle}{\langle z, B_i z\rangle}, \tag{4.52}$$

where $K_i = (\langle u^k, -L_N u^l\rangle_{\mu_N})_{k,l=1}^i \in \mathbb{R}^{i \times i}$ and $B_i = (\langle u^k, u^l\rangle_{\mu_N})_{k,l=1}^i \in \mathbb{R}^{i \times i}$. But, as a consequence of (4.44), we have that, for any $z \in \mathbb{R}^i \setminus \{0\}$,

$$\frac{\langle z, K_i z\rangle}{\|z\|^2} \leq \lambda_{\max}(K_i) \leq \operatorname{tr}(K_i) \leq \frac{\operatorname{cap}(X_i, \mathcal{M}_{i-1}^n)}{\|h_{X_i, \mathcal{M}_{i-1}^n}\|_{2,\mu_N}^2} \left(1 + K e^{-\delta N}\right). \tag{4.53}$$

On the other hand, $\langle z, B_i z\rangle \geq \lambda_{\min}(B_i) \|z\|^2$. In order to obtain a lower bound for $\lambda_{\min}(B_i)$, we exploit again that Gershgorin's Theorem yields

$$\lambda_{\min}(B_i) \geq \min_{j=2,\ldots,i} \left(1 - \frac{\|u^j\|_{1,\mu_N}}{\|u^j\|_{2,\mu_N}} - \sum_{\substack{k=2 \\ k \neq j}}^i \frac{\langle u^j, u^k\rangle}{\|u^j\|_{2,\mu_N} \|u^k\|_{2,\mu_N}}\right)$$

$$\geq \min_{j=2,\ldots,i} \left(1 - \frac{\|h_{X_j, \mathcal{M}_{j-1}^n}\|_{1,\mu_N}}{\sqrt{\mu_N[X_j]}} - \sum_{\substack{k=2 \\ k \neq j}}^i \frac{\langle h_{X_j, \mathcal{M}_{j-1}^n}, h_{X_k, \mathcal{M}_{k-1}^n}\rangle_{\mu_N}}{\sqrt{\mu_N[X_j] \mu_N[X_k]}}\right). \tag{4.54}$$

Thanks to Lemma 3.8 together with the minimal non-degeneracy condition, we get that, for all $j = 2, \ldots, K$, $\mu_N[X_j] < e^{-\delta N} \mu_N[X_1]$ and

$$\|h_{X_j, \mathcal{M}_{j-1}^n}(\sigma)\|_{1,\mu_N} \leq \sum_{M \in \mathcal{M}^n} \sum_{\sigma \in \mathcal{V}^n(M)} \mu_N(\sigma) h_{X_j, \mathcal{M}_{j-1}^n}(\sigma)$$

$$\leq N^{n \cdot q} \mu_N[X_j] \left(1 + e^{-(\mathfrak{C} - \varepsilon(n)c_1)N} + e^{-(\min\{\delta, \mathfrak{C} - \delta\} - \varepsilon(n)c_1)N}\right). \tag{4.55}$$

4.3. Characterization of small eigenvalues

Since $\|h_{X_j,\mathcal{M}_{j-1}^n}\|_{2,\mu_N}^2 \le \|h_{X_j,\mathcal{M}_{j-1}^n}\|_{1,\mu_N}$, the estimates above combined with the Cauchy-Schwarz inequality implies that $\lambda_{\min}(B_i) \ge 1 - N^{n \cdot q}\, e^{-\delta N/2}$ and the assertion follows immediately. □

After having established an upper bound on the small eigenvalues of $-L_N$ that involves only the capacity between metastable sets and the corresponding equilibrium potential, our next task is to prove lower bounds that coincide in the limit $N \to \infty$. In order to apply the Theorem of Kato and Temple, we start with deriving an upper bound on the residual.

Lemma 4.14. Let $A = X_i$ and $B = \mathcal{M}_{i-1}^n$ for $i = 2, \ldots, K$. Further, set

$$\xi = \frac{\langle h_{A,B}, -L_N\, h_{A,B}\rangle_{\mu_N}}{\|h_{A,B}\|_{2,\mu_N}^2}. \tag{4.56}$$

and suppose that $c_1 > 1 + \alpha^{-1}$ is chosen in such a way that $e^{-c_2 N} < \mathbb{P}_{\mu_A}[\tau_B < \tau_A]$. Then,

$$\frac{\|(L_N + \xi)h_{A,B}\|_{2,\mu_N}}{\|h_{A,B}\|_{2,\mu_N}} \le \xi\, e^{\varepsilon(n)c_1 N}. \tag{4.57}$$

Proof. Recall the properties of the equilibrium potential, $h_{A,B}$, and the equilibrium measure, $e_{A,B}$, i.e. $h_{A,B}$ is harmonic in $\sigma \in (A \cup B)^c$ and $e_{A,B}(\sigma) = -(L_N h_{A,B})(\sigma) = -e_{B,A}(\sigma)$ for $\sigma \in A \cup B$. In particular, we have that

$$(L_N + \xi)h_{A,B}(\sigma) = \begin{cases} \xi\, h_{A,B}(\sigma), & \sigma \notin A \cup B, \\ -e_{A,B}(\sigma) + \xi, & \sigma \in A, \\ e_{B,A}(\sigma), & \sigma \in B. \end{cases} \tag{4.58}$$

Hence,

$$\|(L_N + \xi)h_{A,B}\|_{2,\mu_N}^2$$
$$= \sum_{\sigma \in A} \mu_N(\sigma)\,(e_{A,B}(\sigma) - \xi)^2 + \xi^2 \sum_{\sigma \notin A \cup B} \mu_N(\sigma)\, h_{A,B}^2 + \sum_{\sigma \in B} \mu_N(\sigma)\, e_{B,A}(\sigma)^2$$
$$= \xi^2 \|h_{A,B}\|_{2,\mu_N}^2 - 2\xi\, \mathrm{cap}(A,B) + \sum_{\sigma \in A} \mu_N(\sigma)\, e_{A,B}(\sigma)^2 + \sum_{\sigma \in B} \mu_N(\sigma)\, e_{B,A}(\sigma)^2$$
$$= \xi \|h_{A,B}\|_{2,\mu_N}^2 \left(\mathbb{P}_{\nu_{A,B}}[\tau_B < \tau_A] + \mathbb{P}_{\nu_{B,A}}[\tau_A < \tau_B] - \xi \right). \tag{4.59}$$

In view of (2.27), we have that for all $\sigma, \eta \in A$

$$\mathbb{P}_\sigma[\tau_B < \tau_A] \le e^{\varepsilon(n)c_1 N}\, \mathbb{P}_\eta[\tau_B < \tau_A] + e^{-c_2 N}. \tag{4.60}$$

By multiplying both sides with $\mu_N(\eta)$ and summing over all $\eta \in A$, we obtain that

$$\mathbb{P}_\sigma[\tau_B < \tau_A] \leq e^{\varepsilon(n)c_1 N} \frac{\mathrm{cap}(A,B)}{\mu_N[A]} \left(1 + \frac{e^{-c_2 N}}{\mathbb{P}_{\mu_A}[\tau_B < \tau_A]}\right)$$

$$\leq 2e^{\varepsilon(n)c_1 N} \frac{\mathrm{cap}(A,B)}{\mu_N[A]}, \qquad (4.61)$$

where we used the assumption $e^{-c_2 N} < \mathbb{P}_{\mu_A}[\tau_B < \tau_A]$ in the last step. A similar computation reveals that

$$\mathbb{P}_\sigma[\tau_A < \tau_B] \leq e^{\varepsilon(n)c_1 N} \frac{\mathrm{cap}(A,B)}{\mu_N[A]} \left(\frac{\mu_N[A]}{\mu_N[B]} + \frac{e^{-c_2 N}}{\mathbb{P}_{\mu_A}[\tau_B < \tau_A]}\right)$$

$$\leq 2e^{\varepsilon(n)c_1 N} \frac{\mathrm{cap}(A,B)}{\mu_N[A]}. \qquad (4.62)$$

In the last step we took into account that the assumption $\mathcal{M}^n(A) \neq \emptyset$ implies that there exists $M \in \mathcal{M}^n \setminus A$ such that $\mu_N[A] < \mu_N[M]$. Note that (4.61) and (4.62) hold uniformly for all $\sigma \in B$. Plugging (4.61) and (4.62) into (4.59) and using that $\xi = \mathrm{cap}(A,B)/\|h_{A,B}\|^2_{2,\mu_N}$, we get

$$\frac{\|(L_N + \xi)h_{A,B}\|^2_{2,\mu_N}}{\|h_{A,B}\|^2_{2,\mu_N}} \leq \xi^2 \left(4e^{\varepsilon(n)c_1 N} \frac{\|h_{A,B}\|^2_{2,\mu_N}}{\mu_N[A]} - 1\right) \leq \xi^2 e^{2\varepsilon(n)c_1 N}, \quad (4.63)$$

where we used in the last step that, in view of Lemma 3.8, $\|h_{A,B}\|^2_{2,\mu_N} \leq K N^{n \cdot q} \mu_N[A]$ while the number of metastable sets in $\mathcal{M}^n$ grows at most sub-exponential in N. This concludes the proof. $\square$

Lemma 4.15. *Under the assumptions of Theorem 4.12 it holds that*

$$\lambda_i \geq \frac{\mathrm{cap}(X_i, \mathcal{M}^n_{i-1})}{\|h_{X_i, \mathcal{M}^n_{i-1}}\|^2_{2,\mu_N}} \left(1 - e^{-c_{14} N}\right), \qquad \forall\, i = 2, \ldots, K. \qquad (4.64)$$

Proof. In order to shorten notation, we set

$$\xi_i = \frac{\langle h_{X_i, \mathcal{M}^n_{i-1}}, -L_N h_{X_i, \mathcal{M}^n_{i-1}}\rangle_{\mu_N}}{\|h_{X_i, \mathcal{M}^n_{i-1}}\|^2_{2,\mu_N}} = \frac{\mathrm{cap}(X_i, \mathcal{M}^n_{i-1})}{\|h_{X_i, \mathcal{M}^n_{i-1}}\|^2_{2,\mu_N}}.$$

Due to the fact that the assertion (4.64) is trivial if $\lambda_i \geq \xi_i$, let us assume that $\lambda_i < \xi_i$ for all $i = 2, \ldots, K$.

We start with proving a lower bound for λ_K. In order to take advantage of the upper bound in the Theorem of Kato and Temple, we have to find an interval (a_K, b_K) such that $(a_K, b_K) \cap \mathrm{spec}(-L_N) = \emptyset$, see [104, Lemma 3.2]. Since $\lambda_K < \xi_K$ we set $a_K = \xi_K$. Recall that $\lambda_{K+1} \geq e^{-(\delta + 2\varepsilon(n))N} \eta^M$, where η^M is the principal eigenvalue

4.3. Characterization of small eigenvalues

of $-G^n$ subject to zero boundary conditions on $M = \varrho^n(\mathcal{M}^n)$. In view of (4.37) we have that
$$\eta^M \geq \frac{e^{-cN}}{N^{n \cdot q}} \min_{X \in \mathcal{S}_N^n \backslash \mathcal{M}^n} \mathbb{P}_{\mu_X}\left[\tau_{\mathcal{M}^n} < \tau_X\right] \geq \frac{e^{(\mathfrak{C}-c)N}}{N^{n \cdot q}} \xi_K.$$
Now, if we set $b_K = e^{-(\mathfrak{c}+\delta+2\varepsilon(n))N}/N^{n \cdot q} \min_{X \in \mathcal{S}_N^n \backslash \mathcal{M}^n} \mathbb{P}_{\mu_X}\left[\tau_{\mathcal{M}^n} < \tau_X\right]$, the interval (a_K, b_K) contains no eigenvalue of $-L_N$. Hence, the Theorem of Kato and Temple combined with Lemma 4.14 implies that
$$\frac{\xi_K - \lambda_K}{\xi_K} \leq e^{2\varepsilon(n)c_1 N} \frac{\xi_K}{b_K - \xi_K} \leq 2 N^{n \cdot q} e^{-(\mathfrak{C}-\mathfrak{c}-\delta-2\varepsilon(n)(1+c_1))N}, \tag{4.65}$$
and, for N large enough, (4.65) is equivalent to
$$\lambda_K \geq \frac{\operatorname{cap}(X_K, \mathcal{M}_{K-1}^n)}{\|h_{X_K, \mathcal{M}_{K-1}^n}\|_{2,\mu_N}^2} \left(1 - e^{-c_{14} N}\right). \tag{4.66}$$
We prove the lower bound for the remaining eigenvalues $\lambda_{K-1}, \ldots, \lambda_2$ inductively. Suppose that we have already shown that $\lambda_{i+1} \geq \xi_{i+1}\left(1 - e^{-c_{14} N}\right)$. Under the assumption that $\lambda_i < \xi_i$, we set $a_i = \xi_i$ and $b_i = \frac{1}{2}\xi_{i+1}$. As an immediate consequence of the minimal non-degeneracy condition, we have that
$$\mathbb{P}_{\mu_{X_{i+1}}}\left[\tau_{\mathcal{M}_i^n} < \tau_{X_{i+1}}\right] \geq \mathbb{P}_{\mu_{X_{i+1}}}\left[\tau_{\mathcal{M}_{i-1}^n} < \tau_{X_{i+1}}\right] \geq e^{\delta N} \mathbb{P}_{\mu_{X_i}}\left[\tau_{\mathcal{M}_{i-1}^n} < \tau_{X_i}\right]. \tag{4.67}$$
Hence, $\lambda_{i+1} \geq \frac{1}{2}\xi_{i+1} \geq \frac{1}{2} e^{\delta N}/N^{n \cdot q} \xi_i$ which shows that $(a_i, b_i) \cap \operatorname{spec}(-L_N) = \emptyset$ for sufficiently large N. A further application of the Theorem of Kato and Temple combined with (4.57) reveals
$$\frac{\xi_i - \lambda_i}{\xi_i} \leq e^{2\varepsilon(n)c_1 N} \frac{\xi_i}{b_i - \xi_i} \leq 4 N^{n \cdot q} e^{-(\delta - 2\varepsilon(n)c_1)N}. \tag{4.68}$$
This concludes the proof. □

Proof of Theorem 4.12. From (4.50) and (4.64), (4.48) is immediate. Further, in view of Lemma 3.8, we obtain for any $i = 2, \ldots, K$
$$\|h_{X_i, \mathcal{M}_{i-1}^n}\|_{2,\mu_N}^2$$
$$= \sum_{\sigma \in \mathcal{S}_N} \mu_N(\sigma) h_{X_i, \mathcal{M}_{i-1}^n}(\sigma) - \sum_{\sigma \in \mathcal{S}_N} \mu_N(\sigma) h_{X_i, \mathcal{M}_{i-1}^n}(\sigma) h_{\mathcal{M}_{i-1}^n, X_i}(\sigma)$$
$$\geq \sum_{\sigma \in \mathcal{S}_N} \mu_N(\sigma) h_{X_i, \mathcal{M}_{i-1}^n}(\sigma) \left(1 - K N^{n \cdot q} e^{-c_{14} N}\right). \tag{4.69}$$
Since $\|h_{X_i, \mathcal{M}_{i-1}^n}\|_{1,\mu_N} \geq \|h_{X_i, \mathcal{M}_{i-1}^n}\|_{2,\mu_N}^2$, the estimate above yields that
$$\lambda_i = \frac{1}{\mathbb{E}_{\nu_{X_i, \mathcal{M}_{i-1}^n}}[\tau_{\mathcal{M}_{i-1}^n}]} \left(1 - K N^{n \cdot q} e^{-c_{14} N}\right) \tag{4.70}$$
The pointwise estimates in (4.49) follows from Theorem 2.12. □

Chapter 5

Random field Curie–Weiss–Potts model

The present chapter is devoted to the application of the theory developed in the previous chapters. One particular class of models, we are interested in, are *disordered mean field spin systems*. In general, despite of their lack of physical significance, mean field models serve as a good starting point for understanding the phenomena that occur in disordered systems. As an example, we consider the q-spin Potts model on a complete graph, also known as Curie–Weiss–Potts model, in random fields. From a static point of view, it is one of the simplest and less disordered model.

This chapter has two main objectives. First, we are interested in the metastable behavior of the Curie–Weiss–Potts model in the case when the temperature is finite and the distribution of the random field is continuous. We have already mentioned that the entropy plays a particular role in this kind of models, but an exact reduction to a low-dimensional model via lumping techniques is not possible. However, on a heuristic level it is believed that the dynamical behavior on metastable time scales is well described by a diffusion in a mesoscopic potential landscape. For this reason, we construct a family of mesoscopic variables in such a way that the dynamics induced by them is well approximated by a family of Markov processes which reproduce asymptotically the metastable behavior of the original dynamics. Our aim is to compute precisely the mean exit times from a metastable set and to prove the convergence of normalized metastable exit times to an exponential distribution.

In the Sections 1.6 and 1.7 we have seen that variational principles offer a convenient way to derive upper and lower bounds for capacities and, hence, for mean hitting times. Beyond a particular interest in the Curie–Weiss–Potts model, our second objective is to demonstrate how a suitable test function and a non-negative unit flow can be constructed such that the resulting bounds for the capacities coincide asymptotically.

The remainder of this chapter is organized as follows. In Section 5.1, we start with introducing the model and describing the mesoscopic approximation that allows us to deal with the entropy. Afterwards, we present our results. In order to derive sharp estimates for mean hitting times of metastable sets, a first important ingredient is a precise understanding of the measure that is given by the push forward of the random Gibbs measure under the mesoscopic variables. In Section 5.2, we show that the mesoscopic variables satisfy a sharp large deviation principle. In particular, we study the behavior of the corresponding rate function in a neighborhood of critical points. The core of the present chapter are the Sections 5.3 and 5.4, where we construct upper and lower bounds for capacities. In Section 5.3 we start with constructing a test function that is almost harmonic in a small neighborhood of relevant saddle points of the mesoscopic free energy landscape. In view of the Dirichlet principle, this test function is used to derive upper bounds on capacities. As we will see in Section 5.4, this test function is also used to construct in a first step a mesoscopic unit flow. In a second step we construct for each mesoscopic path a subordinate microscopic unit flow and show certain concentration properties along microscopic paths. By exploiting the Berman-Konsowa principle, we obtain in this way lower bounds on capacities. In the finial Section 5.5, the precise computation of mean hitting times is presented.

5.1. Introduction

5.1.1. The model. Let $h = \{h^i\}_{i \in \mathbb{N}}$ be a sequence of independent and identical distributed random variables defined on some probability space, say $(\Omega, \mathcal{B}, \mathbb{P})$, which takes values in $\mathrm{span}\{1\}^\perp \subset \mathbb{R}^q$, where 1 denotes the vector in $\mathbb{R}^q$ whose components are all 1. For the sake of convenience, we will assume that the distribution, $\mathbb{P}_h = \mathbb{P} \circ (h^i)^{-1}$ has bounded support. Most of the quantities that we are going to define depend on the realization of h.

The Curie–Weiss–Potts model is a generalization of the *Ising model* to q components on a complete graph, say, on N vertices where N will be a large parameter. Since the ferromagnetic interaction among the spin variables is assumed to be of the same strength, the actual structure of the graph becomes irrelevant and it suffices to consider just a labeling $\{1, \ldots, N\}$ of the lattice sites. To each site $i \in \{1, \ldots, N\}$ we associate a spin variable σ_i taking values in $\mathcal{S}_0 := \{1, \ldots, q\}$, called the set of different *colors*. In doing so, the *state space* is given by $\mathcal{S}_N = \mathcal{S}_0^N$. Elements of $\mathcal{S}_N$ are denoted by Greek letters σ, η and will be called *configurations*. Given a realization of h, the *random*

5.1. Introduction

Hamiltonian is defined by

$$H_N(\sigma) := -\frac{1}{N}\sum_{i,j=1}^{N}\delta(\sigma_i,\sigma_j) - \sum_{i=1}^{N}\sum_{r\in\mathcal{S}_0} h_r^i\,\delta(\sigma_i,r), \qquad \sigma\in\mathcal{S}_N, \tag{5.1}$$

where $\delta(\sigma_i,\sigma_j)$, the Kronecker symbol, is equal to 1 when $\sigma_i = \sigma_j$ and zero otherwise.

On the measurable space $(\mathcal{S}_N, \mathcal{B}(\mathcal{S}_N))$, where $\mathcal{S}_N$ is equipped with the product topology, we define the *finite volume Gibbs measure* of this model as the *random probability measure*

$$\mu_N(\sigma) := \frac{\exp\left(-\beta H_N(\sigma)\right)}{Z_N}\, q^{-N}, \tag{5.2}$$

where $\beta \geq 0$ is the *inverse temperature* and Z_N is the normalization constant called the *partition function* which is defined by

$$Z_N := \sum_{\sigma\in\mathcal{S}_N}\exp\left(-\beta H_N(\sigma)\right) q^{-N} \equiv \mathbf{E}_{\sigma[N]}\left[\exp(-\beta H_N)\right]. \tag{5.3}$$

Remark 5.1. In the low temperature limit, $\beta \to \infty$, the Gibbs measure favors those configurations where the Hamiltonian is minimal. In view of (5.1), the first term, describing the interaction among the spins, is minimal if all spins have the same color. The second term in (5.1) represents the interaction of the spin variables and the external random magnetic field. It is minimal if in a configuration σ each spin σ_i takes exactly the color corresponding to the maximal component of the field h^i. There is a competition between the contribution coming from the interaction among the spins and the external field, respectively.

We will study the dynamics of a discrete-time Markov chain $\{\sigma(t)\}_{t\in\mathbb{N}_0}$ on the state space $\mathcal{S}_N$ with transition probabilities denoted by p_N. Its discrete generator, L_N, acts on functions $f: \mathcal{S}_N \to \mathbb{R}$ as

$$(L_N f)(\sigma) = \sum_{\eta\in\mathcal{S}_N} p_N(\sigma,\eta)\left(f(\eta) - f(\sigma)\right). \tag{5.4}$$

We assume that the process is *irreducible* and *reversible* with respect to Gibbs measure μ_N. The chain, we consider, evolves by selecting at each step a vertex $i \in \{1,\ldots,N\}$ uniformly at random, proposing a color $r \in \mathcal{S}_0$ uniformly at random among all colors and accepting this proposal according to the probability given by the Metropolis rule. More explicitly, the transition probabilities p_N are given by

$$p_N(\sigma,\eta) := \frac{1}{Nq}\exp\left(-\beta\left[H_N(\eta) - H_N(\sigma)\right]_+\right)\mathbb{1}_{d_H(\sigma,\eta)=1}, \tag{5.5}$$

where $[x]_+ := \max\{x, 0\}$ and $d_H(\sigma, \eta) = 1$ denotes the Hamming distance between σ and η,

$$p_N(\sigma, \sigma) := 1 - \sum_{\eta \in \mathcal{S}_N} p_N(\sigma, \eta). \tag{5.6}$$

Remark 5.2. Note that we consider a lazy version of the Metropolis dynamics since we allow that a chosen site i can draw randomly the color σ_i as well. The reason behind this choice is that there exist $\alpha > 0$ such that $N p_N(\sigma, \sigma^{i,r}) \geq \alpha$ for all $i \in \{1, \dots, N\}$ and $r \in \mathcal{S}_0$. Here, the configuration $\sigma^{i,r}$ is obtained from σ by replacing the color σ_i at site i by the color r.

5.1.2. Coarse graining and mesoscopic approximation. Since we are interested in studying the model at finite temperatures, the entropy is crucial in the static and dynamic description. For this reason, we will investigate the model on a coarse grained scale. To start with, we write

$$\mathcal{M}_p := \left\{ \nu = (\nu_1, \dots, \nu_q) \in [0,1]^q \,\Big|\, \sum_{k \in \mathcal{S}_0} \nu_k = p \right\} \subset \mathbb{R}^q$$

to denote the simplex in $\mathbb{R}^q$, which may be identified with the set of finite measures on $\mathcal{S}_0$. The relative frequencies of colors appearing in a given configuration σ are recorded by means of the *empirical spin distribution*, $\varrho_N : \mathcal{S}_N \to \Gamma_N \subset \mathcal{M}_1$,

$$\sigma \mapsto \varrho_N(\sigma) := \frac{1}{N} \sum_{i=1}^N \delta_{\sigma_i} \tag{5.7}$$

where δ_x is the point-mass at $x \in \mathbb{R}$. Using that $\delta(\sigma_i, \sigma_j) = \sum_{r \in \mathcal{S}_0} \delta(\sigma_i, r) \delta(r, \sigma_j)$ we can rewrite the Hamiltonian (5.1) in terms of the *macroscopic variable*, $\varrho_N(\sigma)$, as

$$H_N(\sigma) = -N \|\varrho_N(\sigma)\|^2 - \sum_{i=1}^N \langle h^i, e^{\sigma_i} \rangle, \tag{5.8}$$

where $e^{\sigma_i} \in \mathbb{R}^q$ denotes the coordinate vector pointing in direction σ_i and $\langle \cdot, \cdot \rangle$ is the usual Euclidean product in $\mathbb{R}^q$. The *macroscopic variables*, ϱ_N, will act as an *order parameter* of this model. We define its distribution under the random Gibbs measure, μ_N, by the *induced measure*, $\mathcal{Q}_N := \mu_N \circ \varrho_N^{-1}$.

Remark 5.3. Note that there are further ways using different representations of the Potts spins to rewrite the Hamiltonian.

(1) In the *hyper-tetrahedral representation* suggested by Wu [115], one can express

$$\delta(\sigma_i, \sigma_j) = \frac{1}{q} + \frac{q-1}{q} \langle v^{\sigma_i}, v^{\sigma_j} \rangle$$

where v^r, for $r \in \mathcal{S}_0$, are unit vectors, pointing in one of the q different directions, which span a $(q-1)$-dimensional hyper-tetrahedron.

5.1. Introduction

(2) In the *polar representation* suggested by Mittag and Stephan [87], one can write

$$\delta(\sigma_i, \sigma_j) = \frac{1}{q} \sum_{r \in S_0} (z_{\sigma_i} \cdot \bar{z}_{\sigma_j})^{r-1}$$

where $z_r = w^r$ with $w = e^{2\pi i/q}$ which is the qth root of unity.

Let us emphasis that, due to the random field, the induced process $\{\varrho_N(\sigma(t))\}_{t \in \mathbb{N}_0}$ is *not* Markovian, $\mathbb{P}$-a.s.. But, there is a canonical construction of an *effective Markov chain* $\{\varrho_N(t)\}_{t \in \mathbb{N}_0}$ on Γ_N that is reversible with respect to $\mathcal{Q}_N$ having the property that $\varrho_N(t) = \varrho_N(\sigma(t))$ in law whenever the induced process $\{\varrho_N(\sigma(t))\}$ is a Markovian. If the Markov chain $\{\sigma(t)\}$ is started in the reversible distribution μ_N, then a classical result of Burke and Rosenblatt [21] implies that $\{\varrho_N(\sigma(t))\}$ is Markovian if and only if for all $\sigma, \eta \in \mathcal{S}_N$ with $\varrho_N(\sigma) = \varrho_N(\eta)$ it holds that $\sum_{\xi: \varrho_N(\xi)=x} p_N(\sigma, \xi) = \sum_{\xi: \varrho_N(\xi)=x} p_N(\eta, \xi)$ for all $x \in \Gamma_N$. However, this condition is only satisfied if all h^i are the same. Due to the fact that $\mathbb{P}_h$ is assumed to be continuous, it can not be expected that, on the macroscopic level, the Markov chain $\{\varrho(t)\}$ and the non-Markovian image chain $\{\varrho(\sigma(t))\}$ have qualitatively the same long-time behavior than. For this reason, our strategy is to construct a family of *mesoscopic variables* having the property that the corresponding effective Markov chain can be seen as a perturbation of the induced dynamics. This strategy was first used in [6].

Due to the fact that the support of the distribution of h^i is bounded, given a sequence $\varepsilon(n) \downarrow 0$ as $n \uparrow \infty$, we can find, for any n, a partition, $\mathcal{H}^n \equiv \{\mathcal{H}^n_1, \ldots, \mathcal{H}^n_{k_n}\}$, such that

$$\mathrm{supp}(\mathbb{P}_h) = \bigcup_{k=1}^{k_n} \mathcal{H}^n_k, \qquad \mathcal{H}^n_k \cap \mathcal{H}^n_l = \emptyset, \quad \forall k \neq l$$

with $\mathrm{diam}(\mathcal{H}^n_k) \leq \varepsilon(n)$. Notice that the sequence of partitions, $\{\mathcal{H}^n\}_{n \in \mathbb{N}}$, can be constructed in such a way that $\mathcal{H}^{n+1}$ is a refinement of $\mathcal{H}^n$. Hence, each realization of the random field $\{h^i\}_{i \in \mathbb{N}}$ induces a random partition of the set $\{1, \ldots, N\}$ into subsets

$$\Lambda^n_k \equiv \Lambda^n_{k,N} := \{i \in \{1, \ldots, N\} \mid h^i \in \mathcal{H}^n_k\}, \qquad k = 1, \ldots, k_n.$$

Let us introduce a family of maps $\varrho^n : \mathcal{S}_N \to \Gamma^n \subset \mathbb{R}^{k_n \cdot q}$,

$$\sigma \mapsto \varrho^n(\sigma) \equiv \varrho^n_N(\sigma) := \sum_{k=1}^{k_n} e^k \otimes \frac{1}{N} \sum_{i \in \Lambda^n_k} \delta_{\sigma_i} \qquad (5.9)$$

where $e^k \in \mathbb{R}^{k_n}$ denotes a coordinate vector in $\mathbb{R}^{k_n}$. Each map ϱ^n represents an averages of *microscopic variables* over blocks of *mesoscopic sizes* which are decreasing in n. Note that the range of ϱ^n is given by

$$\Gamma^n \equiv \Gamma^n_N := \bigtimes_{k=1}^{k_n} \{\nu \in \tfrac{1}{N} \mathbb{N}^q_0 \mid \sum_{r \in S_0} \nu_r = \pi^n_k\} \subset \bigtimes_{k=1}^{k_n} \mathcal{M}_{\pi^n_k}. \qquad (5.10)$$

Here, for any k, $\pi_k^n \equiv \pi_{k,N}^n := |\Lambda_k^n|/N$ is the relative frequencies that h^i takes values in $\mathcal{H}_k^n$ for a given realization of the random field. Note that the random variables π_k concentrate exponentially in N around their mean values $\mathbb{E}[\pi_k] = \mathbb{P}[h^i \in \mathcal{H}_k^n] =: p_k$.

We adopt the notation to denote vectors $\boldsymbol{x} \in \mathbb{R}^{k_n \cdot q}$ by bold symbols whereas its 'components' x^k for $k = 1, \ldots, k_n$ are elements in $\mathbb{R}^q$. Further, in Kronecker products $x \otimes y \in \mathbb{R}^{k_n \cdot q}$ the first vector x will always be an element in $\mathbb{R}^{k_n}$ while the second one y will be in $\mathbb{R}^q$.

The crucial feature of mean field models is that the Hamiltonian (5.1) can be rewritten as a function of the mesoscopic variables. In order to do so, let

$$\bar{h}^k := \frac{1}{|\Lambda_k^n|} \sum_{i \in \Lambda_k^n} h^i \quad \text{and} \quad \tilde{h}^i \equiv \tilde{h}^{i,k} := h^i - \bar{h}^k, \quad \forall i \in \Lambda_k^n \tag{5.11}$$

for each $k = 1, \ldots, k_n$. Then

$$H_N(\sigma) = -N E(\varrho^n(\sigma)) - \sum_{k=1}^{k_n} \sum_{i \in \Lambda_k^n} \langle \tilde{h}^i, e^{\sigma_i} \rangle, \tag{5.12}$$

where the function $E: \mathbb{R}^{k_n \cdot q} \to \mathbb{R}$, is given by

$$\boldsymbol{x} \mapsto E(\boldsymbol{x}) := \left\| \sum_{k=1}^{k_n} x^k \right\|^2 + \sum_{k=1}^{k_n} \langle \bar{h}^k, x^k \rangle. \tag{5.13}$$

Notice that for a given configuration σ, the sum of the mesoscopic variables equals the corresponding macroscopic variable, i.e. $\varrho_N(\sigma) = \sum_{k=1}^{k_n} \varrho^k(\sigma)$.

Let us define the distribution of ϱ^n under the random Gibbs measure, μ_N, through

$$\mathcal{Q}^n \equiv \mathcal{Q}_N^n := \mu_N \circ (\varrho^n)^{-1}. \tag{5.14}$$

Furthermore, we introduce the *mesoscopic free energy*, F^n, which is defined as $F^n : \mathbb{R}^{k_n \cdot q} \to \mathbb{R}$,

$$\boldsymbol{x} \mapsto F^n(\boldsymbol{x}) := -\left\| \sum_{k=1}^{k_n} x^k \right\|^2 - \sum_{k=1}^{k_n} \langle \bar{h}^k, x^k \rangle + \frac{1}{\beta} \sum_{k=1}^{k_n} \pi_k I_{|\Lambda_k|}(x^k/\pi_k). \tag{5.15}$$

The *entropy* or *action function*, $I_{|\Lambda_k|} : \mathbb{R}^q \to \mathbb{R}$, that appears in the expression of F^n is defined as the the *Legendre–Fenchel transform*

$$I_{|\Lambda_k|}(x) := \sup_{t \in \mathbb{R}^q} \left(\langle x, t \rangle - U_{|\Lambda_k|}(t) \right) \tag{5.16}$$

of the *log-moment generating function* $U_{|\Lambda_k|} : \mathbb{R}^q \to \mathbb{R}$,

$$t \mapsto U_{|\Lambda_k|}(t) := \frac{1}{|\Lambda_k|} \ln \mathbf{E}_{\sigma[|\Lambda_k|]}^{\bar{h}} \left[\exp(|\Lambda_k| \langle t, \varrho_{|\Lambda_k|} \rangle) \right] + \frac{1}{|\Lambda_k|} \ln Z_{|\Lambda_k|}^{\bar{h}}$$

$$= \frac{1}{|\Lambda_k|} \sum_{i \in \Lambda_k} \ln \left(\sum_{r \in S_0} \tfrac{1}{q} \exp(\beta \tilde{h}_r^i + t_r) \right), \tag{5.17}$$

5.1. Introduction

where we include the constant in the definition of $U_{|\Lambda_k|}$ to simplify notation in the later.

The Markov chain on state space Γ^n which is reversible with respect to the measure $\mathcal{Q}^n$ is denoted by $\{\varrho(t)\}_{t \in \mathbb{N}_0}$. Its transition probabilities are given by

$$r^n(\boldsymbol{x}, \boldsymbol{y}) \equiv r_N^n(\boldsymbol{x}, \boldsymbol{y}) := \frac{1}{\mathcal{Q}^n(\boldsymbol{x})} \sum_{\sigma \in \mathcal{S}^n[\boldsymbol{x}]} \mu_N(\sigma) \sum_{\eta \in \mathcal{S}^n[\boldsymbol{y}]} p_N(\sigma, \eta), \qquad \boldsymbol{x}, \boldsymbol{y} \in \Gamma^n, \tag{5.18}$$

where $\mathcal{S}^n[\boldsymbol{x}] := (\varrho^n)^{-1}(\boldsymbol{x})$ is the set-valued preimage of ϱ^n. If n is chosen such that for all $\boldsymbol{x}, \boldsymbol{y} \in \Gamma^n$ with $r^n(\boldsymbol{x}, \boldsymbol{y}) > 0$ it holds that for $\beta \varepsilon(n) \leq 1$, then

$$\max_{\sigma \in \mathcal{S}^n[\boldsymbol{x}], \eta \in \mathcal{S}^n[\boldsymbol{y}]} \left| \frac{p_N(\sigma, \eta) \left| \{ \eta \in \mathcal{S}^n[\boldsymbol{y}] \mid d_H(\sigma, \eta) = 1 \} \right|}{r^n(\boldsymbol{x}, \boldsymbol{y})} - 1 \right| \leq 3\beta \varepsilon(n). \tag{5.19}$$

In other words, the Markov chain $\{\varrho^n(t)\}$ can be seen as a good approximation of the process $\{\varrho^n(\sigma(t))\}$. Finally, we denote by L^n the discrete generator of the Markov chain $\{\varrho^n(t)\}$.

In order to simplify the notation in what follows, we will frequently drop the superscript n, identify $k_n \equiv n$ and refer to the generic partition $\Lambda_1, \ldots, \Lambda_n$.

5.1.3. Main results. In the sequel, we assume that the inverse temperature β and the distribution, $\mathbb{P}_h$, of the random field are such that there exists more than one local minima of F^n. Here, the coarsening parameter $n \ll N$ is arbitrary but fixed. Let $m \in \Gamma^n$ be a local minimum and let us denote by $M \subset \Gamma^n$ the set of deeper local minima of F^n, i.e. let $F^n(m') < F^n(m)$ for all $m' \in M$. Further, we assume that there exists a unique relevant saddle point, $z \in \Gamma^n$, of index one that separates m and M, i.e. $F^n(z) = \min_\gamma \max_{x \in \gamma} F^n(x)$ where the minimum is taken over all mesoscopic paths in Γ^n that connects m and M. To each critical point $\boldsymbol{x} \in \Gamma^n$ of F^n we associate a point $x \in \Gamma_N$ that is given by $x = x(\boldsymbol{x}) = \sum_{k=1}^n x^k$. In Lemma 5.8 we will show that $x(\boldsymbol{x}) \neq y(\boldsymbol{y})$ if the critical points $\boldsymbol{x}$ and $\boldsymbol{y}$ are distinct. Further, we set $A = \mathcal{S}^n[m]$ and $B = \mathcal{S}^n[M]$.

The main result of the present chapter is the following

Theorem 5.4. *With the notations and assumptions introduced above, we have that $\mathbb{P}_h$-almost surely, for all but finitely many N,*

$$Z_N \, \mathrm{cap}(A, B) = \frac{\beta |\bar{\gamma}_1|}{2\pi N} \frac{\exp(-\beta N F_N(z)) \left(1 + o_N(1)\right)}{\sqrt{\left| \det \left(I_q - 2\beta \left(D(z) - \mathbb{E}_h \left[u^1(2\beta z) \cdot u^1(2\beta z)^{\mathrm{T}} \right] \right) \right) \right|}}, \tag{5.20}$$

where $z = z(z)$, $D(z) = \text{diag}(z_1, \ldots, z_q)$ and $u^i(2\beta z) \in \mathbb{R}^q$ is defined componentwise by

$$u_r^i(t) := \frac{\exp(t_r + \beta h_r^i)}{\sum_{s \in \mathcal{S}_0} \exp(t_s + \beta h_s^i)}, \qquad \forall k \in \mathcal{S}_0. \tag{5.21}$$

Further, $\bar{\gamma}_1$ is the unique negative solution of the explicit equation that is given in (5.87), and

$$F_N(z) = \|z\|^2 - \frac{1}{\beta N} \sum_{i=1}^{N} \ln\left(\sum_{r \in \mathcal{S}_0} \tfrac{1}{q} \exp(2\beta z_r + \beta h_r^i)\right). \tag{5.22}$$

For the mean exit times from A and its distribution in the random field Curie–Weiss–Potts model we prove

Theorem 5.5. *Let us consider the notation introduced above and in Theorem 5.4. Further, we assume that there exists $\delta > 0$ such that, for large enough N, $\delta < F^n(z) - F^n(m)$ and there is no local minimum m' of F^n with $F^n(m') \in [F^n(m), F^n(m) + \delta]$. Then,*

(i) *$\mathbb{P}_h$-almost surely, for all but finitely many values of N, we have that*

$$\mathbb{E}_{\nu_{A,B}}[\tau_B] = \frac{2\pi N}{\beta |\bar{\gamma}_1|} K(\beta, m, z) \exp\left(\beta N \left(F_N(z) - F_N(m)\right)\right)(1 + \mathcal{O}_N(1)) \tag{5.23}$$

where

$$K(\beta, m, z) = \sqrt{\frac{\left|\det\left(I_q - 2\beta\left(D(z) - \mathbb{E}_h\left[u^1(2\beta z) \cdot u^1(2\beta z)^{\mathrm{T}}\right]\right)\right)\right|}{\det\left(I_q - 2\beta\left(D(m) - \mathbb{E}_h\left[u^1(2\beta m) \cdot u^1(2\beta m)^{\mathrm{T}}\right]\right)\right)}}. \tag{5.24}$$

(ii) *provided that the coarsening level n, used in the definition of A, is chosen large enough to ensure that $\varepsilon(n)c_1 < (F_N(z) - F_N(m))/4$ for a sufficiently large $c_1 > 1 + \alpha^{-1}$, then*

$$\mathbb{E}_\sigma[\tau_B] = \mathbb{E}_{\nu_{A,B}}[\tau_B](1 + \mathcal{O}_N(1)), \qquad \forall \sigma \in A. \tag{5.25}$$

(iii) *under the same assumption as in (ii), for all $t \in \mathbb{R}_+$ and $\sigma \in A$*

$$\mathbb{P}_\sigma[\tau_B / \mathbb{E}_\sigma[\tau_B] > t] \longrightarrow e^{-t}, \qquad \text{as} \quad N \to \infty. \tag{5.26}$$

5.2. Induced measure and free energy landscape

The first important ingredient to derive sharp estimates for mean hitting times is a precise understanding of the induced measure $\mathcal{Q}^n$. The infinite-volume equilibrium (Gibbs) measures of the Curie-Weiss model of Ising ($q = 2$) or Potts ($q \geq 3$) type in the presents of a random field and the particular influence of the disorder on the typical behavior of Gibbs measure in large but finite volumes was analyzed in detail in [1, 79, 70]. In this section, our first aim is to show that the mesoscopic variable, ϱ^n, satisfies a *sharp large deviation principle* under the Gibbs measure with rate function F^n. Next,

5.2. Induced measure and free energy landscape

we study the behavior of the induced random measure $\mathcal{Q}^n$ in the neighborhood of critical points of F^n.

5.2.1. Sharp large deviation principle.
Let us begin by writing for $x \in \Gamma^n$

$$Z_N \mathcal{Q}^n(x) = \exp(\beta N E(x)) \prod_{k=1}^n \sum_{\sigma \in \mathcal{S}_{|\Lambda_k|}} \mathbb{1}_{\varrho^k(\sigma) = x^k} \exp\left(\beta \sum_{i \in \Lambda_k} \langle \tilde{h}^i, e^{\sigma_i} \rangle\right) q^{-|\Lambda_k|}$$

$$=: \exp(\beta N E(x)) \prod_{k=1}^n Z^{\tilde{h}}_{|\Lambda_k|} \mathbf{E}^{\tilde{h}}_{\sigma[|\Lambda_k|]}\left[\mathbb{1}_{\varrho_{|\Lambda_k|} = x^k / \pi_k}\right], \tag{5.27}$$

where $Z^{\tilde{h}}_{|\Lambda_k|}$ is the normalization constant of the $\tilde{h}$-tilted measure. It remains to control the probability with respect to the $\tilde{h}$-tilted measure of the event, $\{\varrho_{|\Lambda_k|} = x\}$, that the macroscopic variable $\varrho_{|\Lambda_k|}$ takes a given value $x \in \Gamma_{|\Lambda_k|} \subset \mathcal{M}_1$ in the limit when $N \to \infty$.

For this purpose, we begin with recalling some well-known properties of the log-moment generating function, $U_{|\Lambda_k|}$, and its Legendre–Fenchel transform, $I_{|\Lambda_k|}$, taken from [46, 103]. As a consequence of Hölder's inequality, $U_{|\Lambda_k|}$ is a convex function on $\mathbb{R}^q$. This implies, that $I_{|\Lambda_k|}$ is a proper convex function and lower semi-continuous. Moreover, $U_{|\Lambda_k|}$ is infinitely differentiable on $\mathbb{R}^q$ and the range of its gradient map coincides with the *relative interior* of $\mathcal{M}_1$ that we denote by $\mathrm{ri}(\mathcal{M}_1)$.

We recall that for a convex subset $C \subseteq \mathbb{R}^q$ the *affine hull* of C, $\mathrm{aff}(C)$, is defined to be the intersection of all affine sets which contain C. This concept allows us to define $\mathrm{ri}(C)$ as the set of all $x \in C$ for which there exists an $\varepsilon > 0$ such that $B_\varepsilon(x) \cap \mathrm{aff}(C) \subset C$. Here $B_\varepsilon(x)$ denotes an open ball in $\mathbb{R}^q$ around x of radius ε.

As an application of [46, Theorem VI.5.7],

$$\mathrm{ri}(\mathrm{dom}(I_{|\Lambda_k|})) = \mathrm{ri}(\mathcal{M}_1) \subseteq \mathrm{dom}(I_{|\Lambda_k|}) \subseteq \mathcal{M}_1,$$

while for all $x \in \mathbb{R}^q \setminus \mathcal{M}_1$ the entropy $I_{|\Lambda_k|}(x) = \infty$. Since $U_{|\Lambda_k|}$ is smooth, $I_{|\Lambda_k|}$ is strictly convex on every convex subset of $\mathrm{ri}(\mathcal{M}_1)$. Further, notice that for all $r \in \mathcal{S}_0$

$$\frac{\partial U_{|\Lambda_k|}}{\partial t_r}(t) = \frac{1}{|\Lambda_k|} \sum_{i \in \Lambda_k} \frac{\exp(t_r + \beta \tilde{h}^i_r)}{\sum_{s \in \mathcal{S}_0} \exp(t_s + \beta \tilde{h}^i_s)} =: \frac{1}{|\Lambda_k|} \sum_{i \in \Lambda_k} u^{i,k}_r(t). \tag{5.28}$$

Thus, the gradient map $\nabla U_{|\Lambda_k|}$ is *not* one-to-one on $\mathbb{R}^q$ because for any $x \in \mathrm{ri}(\mathcal{M}_1)$ and $t^* \equiv t^*(x) \in \mathbb{R}^q$ such that $x = \nabla U_{|\Lambda_k|}(t^*)$, the one-parameter family $t^*_\gamma := t^* + \gamma \mathbb{1}$ satisfies as well $x = \nabla U_{|\Lambda_k|}(t^*_\gamma)$ for all $\gamma \in \mathbb{R}$. Nevertheless, since

$$\langle x, t^* \rangle - U_{|\Lambda_k|}(t^*) = \langle x, t^*_\gamma \rangle - U_{|\Lambda_k|}(t^*_\gamma) \qquad \forall \gamma \in \mathbb{R},$$

the entropy is well defined, see also [103, Theorem 26.4], and given by the formula

$$I_{|\Lambda_k|}(x) = \langle x, t^* \rangle - U_{|\Lambda_k|}(t^*). \tag{5.29}$$

Note that, due to [103, Theorem 23.5], it holds that

$$t^* \in \partial I_{|\Lambda_k|}(x) \iff x = \nabla U_{|\Lambda_k|}(t^*) \tag{5.30}$$

where $\partial I_{|\Lambda_k|}(x)$ denotes the *subdifferential of* $I_{|\Lambda_k|}$ at x which is defined through

$$\partial I_{|\Lambda_k|}(x) := \{t \in \mathbb{R}^q \mid I_{|\Lambda_k|}(z) \geq I_{|\Lambda_k|}(x) + \langle t, z - x \rangle, \, \forall z \in \mathbb{R}^q \}.$$

For obvious reasons, $\partial I_{|\Lambda_k|}(x)$ is an affine subspace of $\mathbb{R}^q$ with difference space V_0, where we decompose $\mathbb{R}^q$ into a direct sum of

$$V_0 = \text{span}\{1\} \quad \text{and} \quad V_1 = V_0^\perp. \tag{5.31}$$

Moreover, a simple computation shows, see Lemma 5.6, that the log-moment generating function $U_{|\Lambda_k|}(t)$ is strict convex in the direction orthogonal the V_0 and the leading principal minor, which we denote by $\left[\text{Hess}\, U_{|\Lambda_k|}(t)\right]_q$, is positive definite. This implies that the entropy $I_{|\Lambda_k|}$ with respect to aff($\mathcal{M}_1$) is *essentially smooth*, i.e.

(a) $I_{|\Lambda_k|}$ is differentiable throughout ri($\mathcal{M}_1$) and
(b) for any sequence $\{x^i\}$ in ri($\mathcal{M}_1$) which converge to a boundary point $x \in \mathcal{M}_1$, $\|\nabla I_{|\Lambda_k|}(x^i)\| \to \infty$.

Lemma 5.6. *For* $t \in \mathbb{R}^q$ *set* $x = \nabla U_{|\Lambda_k|}(t) \in \mathcal{M}_1$. *Then the matrix* $\text{Hess}\, U_{|\Lambda_k|}(t)$ *has exactly one eigenvalue zero in the direction* $1 \in \mathbb{R}^q$. *Moreover, there exists* $c \equiv c(t) > 0$ *such that the eigenvalues corresponding to* $\text{span}\{1\}^\perp$, *as well as, the eigenvalues of* $\left[\text{Hess}\, U_{|\Lambda_k|}(t)\right]_q$ *are bounded from below by* c.

Proof. In view of (5.28), the symmetric matrix $B \equiv \text{Hess}\, U_{|\Lambda_k|}(t)$ can be written as the difference between a diagonal matrix and a sum of rang-1 matrices, i.e.

$$B = \frac{1}{|\Lambda_k|} \sum_{i \in \Lambda_k} \text{diag}(u_1^{i,k}, \ldots, u_q^{i,k}) - u^{i,k} \cdot (u^{i,k})^T$$

$$= \text{diag}(x_1, \ldots, x_q) - \frac{1}{|\Lambda_k|} \sum_{i \in \Lambda_k} u^{i,k} \cdot (u^{i,k})^T$$

where $u^{i,k} \equiv u^{i,k}(t^*)$. Since $U_{|\Lambda_k|}$ is a convex function, the matrix B is positive semi-definite. Moreover, it is easy to check that 1 is an eigenvector to the eigenvalue 0. Let us order the eigenvalues of B such that $\lambda_1(B) \leq \lambda_2(B) \leq \ldots \leq \lambda_q(B)$. Using the interlacing property of symmetric matrices, we get

$$0 = \lambda_1(B) \leq \lambda_1(A) \leq \lambda_2(B) \leq \ldots \leq \lambda_{q-1}(B) \leq \lambda_{q-1}(A) \leq \lambda_q(B) \tag{5.32}$$

where $A \equiv \left[\text{Hess}\, U_{|\Lambda_k|}(t)\right]_q$. By applying Gershgorin's Theorem, we obtain that

$$\text{spec}(A) \subset \bigcup_{r \in S_0} [A_{rr} - R_r, A_{rr} + R_r]$$

5.2. Induced measure and free energy landscape

where

$$R_r := \sum_{\substack{s=1 \\ s \neq r}}^{q-1} |A_{rs}| = \frac{1}{|\Lambda_k|} \sum_{i \in \Lambda_k} u_r^{i,k} \sum_{\substack{s=1 \\ s \neq r}}^{q-1} u_s^{i,k} = x_r - \frac{1}{|\Lambda_k|} \sum_{i \in \Lambda_k} u_r^{i,k} \left(u_q^{i,k} + u_r^{i,k} \right).$$

Since $\mathrm{diam}(\mathcal{H}_k) \leq \varepsilon(n)$ and by (5.28), there exists a $\delta(t) > 0$ such that the minimal component of x is bounded from below by $\delta(t)$. Hence,

$$A_{rr} - R_r = \frac{1}{|\Lambda_k|} \sum_{i \in \Lambda_k} u_r^{i,k} u_q^{i,k}$$

$$= \frac{1}{|\Lambda_k|^2} \sum_{i,j \in \Lambda_k} u_r^{i,k} u_q^{j,k} \frac{u_q^{i,k}}{u_q^{j,k}} \geq x_r x_q \, e^{-2\beta\varepsilon(n)} \geq c > 0$$

where we set $c := \delta(t)^2 \, e^{-2\beta\varepsilon(n)}$. $\square$

Now, we can state the sharp large deviation estimate for the h-tilted expectation.

Proposition 5.7. *Let $y \in \mathrm{ri}(\mathcal{M}_1)$ and consider a sequence $\{x^{|\Lambda_k|}\}$ such that $x^{|\Lambda_k|} \in \Gamma_{|\Lambda_k|}$ and $\|x^{|\Lambda_k|} - y\|_2 < \frac{1}{|\Lambda_k|}$. Further define $t^{|\Lambda_k|} \in V_1 \subset \mathbb{R}^q$ through $x^{|\Lambda_k|} = \nabla U_{|\Lambda_k|}(t^{|\Lambda_k|})$ if such a $t^{|\Lambda_k|}$ exists while otherwise $\|t^{|\Lambda_k|}\|_2 = \infty$. Then, for all but finitely many N,*

$$Z_{|\Lambda_k|}^{\tilde{h}} \, \mathbf{E}_{\sigma[|\Lambda_k|]}^{\tilde{h}} \left[\mathbb{1}_{\varrho_{|\Lambda_k|} = x^{|\Lambda_k|}} \right] = \frac{\exp\left(-|\Lambda_k| I_{|\Lambda_k|}(x^{|\Lambda_k|})\right)}{\sqrt{(2\pi|\Lambda_k|)^{q-1} \det\left[\mathrm{Hess}\, U_{|\Lambda_k|}(t^{|\Lambda_k|})\right]_q}} (1 + o_N(1)),$$
(5.33)

where $o_N(1) \to 0$ as $N \to \infty$.

Proof. Sharp large deviation estimates were obtained in [26, 39] for random vectors and in the Ising model. The proof of (5.33) modifies their strategy to the case, where the Hessian is degenerate.

Recall that, $\mathbb{P}$-a.s., $|\Lambda_k| = p_k(1 + o_N(1)) N$. Now, consider for any $t \in \mathbb{R}^q$ and $x \in \Gamma_{|\Lambda_k|}$

$$Z_{|\Lambda_k|}^{\tilde{h}} \, \mathbf{E}_{\sigma[|\Lambda_k|]}^{\tilde{h}} \left[\mathbb{1}_{\varrho_{|\Lambda_k|} = x} \right]$$

$$= e^{-|\Lambda_k|(\langle x, t \rangle - U_{|\Lambda_k|}(t))} \, \mathbf{E}_{\sigma[|\Lambda_k|]}^{\tilde{h}} \left[\mathbb{1}_{\varrho_{|\Lambda_k|} = x} \frac{e^{|\Lambda_k|\langle t, \varrho_{|\Lambda_k|}\rangle}}{\mathbf{E}_{\sigma[|\Lambda_k|]}^{\tilde{h}}\left[e^{|\Lambda_k|\langle t, \varrho_{|\Lambda_k|}\rangle}\right]} \right]$$

$$= e^{-|\Lambda_k|(\langle x, t \rangle - U_{|\Lambda_k|}(t))} \, \widehat{\mathbf{E}}_{\sigma[|\Lambda_k|]}^{t} \left[\mathbb{1}_{\varrho_{|\Lambda_k|} = x} \right].$$

Since $y \in \mathrm{ri}(\mathcal{M}_1)$, there exists $\varepsilon > 0$ such that $B_\varepsilon(y) \cap \mathrm{aff}(\mathcal{M}_1) \subset \mathrm{ri}(\mathcal{M}_1)$. By choosing N large enough, we can ensure that $\|x^{|\Lambda_k|} - y\|_2 < \varepsilon/2$. This implies that $\|t^{|\Lambda_k|}\|_2 < \infty$. Setting $t = t^{|\Lambda_k|}$ and using (5.29), we obtain that

$$\mathbf{E}_{\sigma[|\Lambda_k|]}^{\tilde{h}} \left[\mathbb{1}_{\varrho_{|\Lambda_k|} = x^{|\Lambda_k|}} \right] = e^{-|\Lambda_k| I_{|\Lambda_k|}(x^{|\Lambda_k|})} \, \widehat{\mathbf{E}}_{\sigma[|\Lambda_k|]}^{t^{|\Lambda_k|}} \left[\mathbb{1}_{\varrho_{|\Lambda_k|} = x^{|\Lambda_k|}} \right]. \quad (5.34)$$

On the other hand, for any $x \in \Gamma_{|\Lambda_k|}$, it holds

$$\mathbb{1}_{\varrho_{|\Lambda_k|}(\sigma) = x} = \prod_{r=1}^{q-1} \mathbb{1}_{\varrho_{|\Lambda_k|}(\sigma)_r = x_r} = \frac{1}{(2\pi)^{q-1}} \int_Q e^{i\,|\Lambda_k|\,\langle \varrho_{|\Lambda_k|}(\sigma) - x, \binom{z}{0}\rangle}\, dz,$$

where $Q = [-\pi, \pi]^{q-1}$. Hence,

$$\widehat{\mathbb{E}}_{\sigma[|\Lambda_k|]}^{t^{|\Lambda_k|}}\!\left[\mathbb{1}_{\varrho_{|\Lambda_k|} = x^{|\Lambda_k|}}\right] = \frac{1}{(2\pi)^{q-1}} \int_Q \exp\!\big(|\Lambda_k|\, G_{|\Lambda_k|}(z)\big)\, dz \qquad (5.35)$$

where

$$G_{|\Lambda_k|}(z) = U_{|\Lambda_k|}\!\big(t^{|\Lambda_k|} + \mathrm{i}\binom{z}{0}\big) - U_{|\Lambda_k|}\!\big(t^{|\Lambda_k|}\big) - \mathrm{i}\,\langle x^{|\Lambda_k|}, \binom{z}{0}\rangle, \qquad z \in \mathbb{R}^{q-1}.$$

Now, set $\delta = \pi/4$. For a given $\gamma_{|\Lambda_k|} := \sqrt{K_\varepsilon \ln|\Lambda_k|/|\Lambda_k|}$, where K_ε will be defined below, there exists N_ε such that $\gamma_{|\Lambda_k|} < \delta/2$ and $\|x^{|\Lambda_k|} - y\|_2 < \varepsilon/2$ for all $|\Lambda_k| > N_\varepsilon$. Moreover, we rewrite (5.35) as

$$\widehat{\mathbb{E}}_{\sigma[|\Lambda_k|]}^{t^{|\Lambda_k|}}\!\left[\mathbb{1}_{\varrho_{|\Lambda_k|} = x^{|\Lambda_k|}}\right] = \frac{1}{(2\pi)^{q-1}} \int_Q e^{|\Lambda_k|\, G_{|\Lambda_k|}(z)}\, \mathbb{1}_{\|z\|_2 \leq \gamma_{|\Lambda_k|}}\, dz$$

$$+ \frac{1}{(2\pi)^{q-1}} \int_Q e^{|\Lambda_k|\, G_{|\Lambda_k|}(z)}\, \mathbb{1}_{\|z\|_2 > \gamma_{|\Lambda_k|}}\, dz. \qquad (5.36)$$

We start with studying the asymptotics of the first integral. Since $U_{|\Lambda_k|}$ is holomorphic, the following expansion is valid for all $\|z\|_2 < \delta$

$$G_{|\Lambda_k|}(z) = \mathrm{i}\,\langle \nabla U_{|\Lambda_k|}(t^{|\Lambda_k|}), \binom{z}{0}\rangle - \tfrac{1}{2}\langle z, A z\rangle + R_3\!\big(t^{|\Lambda_k|} + \mathrm{i}\binom{z}{0}\big) - \mathrm{i}\,\langle x^{|\Lambda_k|}, \binom{z}{0}\rangle$$

$$= -\tfrac{1}{2}\langle z, A z\rangle + R_3\!\big(t^{|\Lambda_k|} + \mathrm{i}\binom{z}{0}\big), \qquad (5.37)$$

where $A := \big[\mathrm{Hess}\,U_{|\Lambda_k|}(t^{|\Lambda_k|})\big]_q$ and $R_3\!\big(t^{|\Lambda_k|} + \mathrm{i}\binom{z}{0}\big) = \sum_{|\alpha| \geq 3} a_\alpha^{|\Lambda_k|}(\mathrm{i}z)^\alpha$. Notice that the coefficients in the remainder term can be bounded from above by means of Cauchy's theorem,

$$|a_\alpha^{|\Lambda_k|}| \leq \frac{1}{\delta^\alpha} \sup_{\xi \in \mathrm{cl}(P_\delta(t^{|\Lambda_k|}))} |U_{|\Lambda_k|}(\xi)|. \qquad (5.38)$$

Here, we denote by $P_\delta(t) := \{z \in \mathbb{C}^q \mid \|t - z\|_\infty < \delta\} \subset \mathbb{C}^q$ the polydisc with center $t \in \mathbb{R}^q$ and radius δ. Thus, we are left with the task to show that, for all $\xi \in \mathrm{cl}(P_\delta(t^{|\Lambda_k|}))$, the supremum of $|U_{|\Lambda_k|}(\xi)|$ is uniformly bounded from above for all $|\Lambda_k| > N_\varepsilon$. Since $\delta = \pi/4$, it holds for all $\xi \in \mathrm{cl}(P_\delta(t^{|\Lambda_k|}))$ that $|\Im(U_{|\Lambda_k|}(\xi))| \leq \delta$. Note that for $\xi^1 + \mathrm{i}\xi^2 \in \mathbb{C}^q$,

$$\Re\!\big(U_{|\Lambda_k|}(\xi^1 + \mathrm{i}\xi^2)\big) - U_{|\Lambda_k|}(\xi^1)$$

$$= \frac{1}{2|\Lambda_k|} \sum_{i \in \Lambda_k} \ln\!\Big(\sum_{r,s \in S_0} u_r^{i,k}(\xi^1)\, u_s^{i,k}(\xi^1)\, \cos\!\big(\xi_r^2 - \xi_s^2\big)\Big), \qquad (5.39)$$

5.2. Induced measure and free energy landscape

which implies that $|\Re(U_{|\Lambda_k|}(\xi))| \leq U_{|\Lambda_k|}(t^{|\Lambda_k|}) + \delta$ for all $\xi \in \mathrm{cl}(P_\delta(t^{|\Lambda_k|}))$. Recall that $u^i(t)$ is defined in (5.28). Thus, $|U_{|\Lambda_k|}(\xi)| \leq |\Re(U_{|\Lambda_k|}(\xi))| + |\Im(U_{|\Lambda_k|}(\xi))| \leq U_{|\Lambda_k|}(t^{|\Lambda_k|}) + 2\delta$. It remains to control $U_{|\Lambda_k|}(t^{|\Lambda_k|})$. However, since

$$x^{|\Lambda_k|} = \nabla U_{|\Lambda_k|}(t^{|\Lambda_k|}) = \frac{1}{|\Lambda_k|} \sum_{i \in \Lambda_k} u^{i,k}(t^{|\Lambda_k|}) \leq u^{j,k}(t^{|\Lambda_k|}) e^{2\beta\varepsilon(n)} \quad \forall j \in \Lambda_k.$$

and, by rewriting $U_{|\Lambda_k|}$, we obtain

$$\begin{aligned} U_{|\Lambda_k|}(t^{|\Lambda_k|}) &= -\frac{1}{q} \sum_{r \in S_0} \frac{1}{|\Lambda_k|} \sum_{j \in \Lambda_k} \ln\left(u_r^{j,k}(t^{|\Lambda_k|})\right) \\ &\leq -\frac{1}{q} \sum_{r \in S_0} \ln x_r^{|\Lambda_k|} + 2\beta\varepsilon(n) \leq \ln \frac{2}{\varepsilon} + 2\beta\varepsilon(n), \end{aligned} \quad (5.40)$$

where we used that $t^{|\Lambda_k|} \perp 1$, as well as, $h^i \perp 1$ for all $i \in \Lambda_k$ and that by construction the minimal component of $x^{|\Lambda_k|}$ is at least larger than $\varepsilon/2$. Thus, we can conclude that there exists $M_\varepsilon < \infty$ such that, for all $|\Lambda_k| > N_\varepsilon$, $\sup_{\xi \in \mathrm{cl}(P_\delta(t^{|\Lambda_k|}))} |U_{|\Lambda_k|}(\xi)| \leq M_\varepsilon$.

Since $\|z\|_2 \leq \gamma_{|\Lambda_k|} < \delta/2$ for all $|\Lambda_k| > N_\varepsilon$, the remainder can be bounded from above by

$$\left|R_3\left(t_N^* + \mathrm{i}\binom{z}{0}\right)\right| \leq M_\varepsilon \frac{\|z\|_2^3}{\delta^3} \sum_{k=0}^\infty \left(\frac{\|z\|_2}{\delta}\right)^k \leq 2 M_\varepsilon \frac{\|z\|_2^3}{\delta^3}. \quad (5.41)$$

Hence, (5.37) together with the Lemma 5.6 implies

$$\begin{aligned} \int_{\mathbb{R}^{q-1}} & e^{|\Lambda_k| G_{|\Lambda_k|}(z)} \mathbb{1}_{\|z\|_2 \leq \gamma_{|\Lambda_k|}} \, dz \\ &= |\Lambda_k|^{-\frac{q-1}{2}} \int_{\mathbb{R}^{q-1}} e^{-\frac{1}{2}\langle z, A z\rangle} \mathbb{1}_{\|z\|_2 \leq \sqrt{K_\varepsilon \ln |\Lambda_k|}} \, dz \, (1 + o_N(1)) \\ &= \frac{(2\pi)^{q-1}}{\sqrt{(2\pi|\Lambda_k|)^{q-1} \det\left[\mathrm{Hess}\, U_{|\Lambda_k|}(t^{|\Lambda_k|})\right]_q}} (1 + o_N(1)). \end{aligned} \quad (5.42)$$

In order to complete the proof, it remains to show that the contribution to (5.33) coming from the second integral in (5.36) is negligible, i.e. that the following *locality condition* holds

$$|\Lambda_k|^{\frac{q-1}{2}} \int_Q \left|e^{|\Lambda_k| G_{|\Lambda_k|}(z)}\right| \mathbb{1}_{\|z\|_2 > \gamma_{|\Lambda_k|}} \, dz = o_N(1).$$

By exploiting the representation (5.39), we obtain that

$$\max_{\substack{z \in Q \\ \|z\|_2 > \gamma_{|\Lambda_k|}}} \Re\big(G_{|\Lambda_k|}(z)\big) \leq \frac{1}{2|\Lambda_k|} \sum_{i \in \Lambda_k} \ln\left(1 - \frac{4}{\pi^2} u_q^{i,k}(t^{|\Lambda_k|}) \sum_{r=1}^{q-1} u_r^{i,k}(t^{|\Lambda_k|}) z_r^2\right)$$

$$\leq -\frac{2}{\pi^2} e^{-2\beta\varepsilon(n)} x_q^{|\Lambda_k|} \sum_{r=1}^{q-1} x_r^{|\Lambda_k|} z_r^2$$

$$\leq -\frac{\varepsilon^2}{2\pi^2} e^{-2\beta\varepsilon(n)} \gamma_{|\Lambda_k|}^2. \tag{5.43}$$

In the first step, we use that $1 - \cos x \geq 2x^2/\pi^2$ for all $|x| \leq \pi$. In the second step, we took advantage of (5.28) and the fact that $e^{-2\beta\varepsilon(n)} \leq u_r^{i,k}(t)/u_r^{j,k}(t) \leq e^{2\beta\varepsilon(n)}$ for all $i,j \in \Lambda_k$ and $r \in \mathcal{S}_0$. In the last step we used that by construction the minimal component of $x^{|\Lambda_k|}$ is at least larger than $\varepsilon/2$. Thus, by choosing $K_\varepsilon = \frac{\pi^2}{\varepsilon^2} e^{2\beta\varepsilon(n)} q$, we get

$$|\Lambda_k|^{\frac{q-1}{2}} \int_Q \left| e^{|\Lambda_k| G_{|\Lambda_k|}(z)} \right| \mathbb{1}_{\|z\|_2 > \gamma_{|\Lambda_k|}} \, dz \leq (2\pi)^{q-1} |\Lambda_k|^{-1/2}. \tag{5.44}$$

Thus, by combing (5.42) and (5.44), we conclude the proof. $\square$

For $y \in \mathrm{ri}(\mathcal{M}_1)^{\otimes n}$ and N large enough, let us consider $x \in \Gamma^n$ with $\|x^k/\pi_k - y^k\|_2 < \varepsilon/2$ where $\varepsilon > 0$ is chosen in such a way that $B_\varepsilon(y^k) \cap \mathrm{aff}(\mathcal{M}_1) \subset \mathrm{ri}(\mathcal{M}_1)$ for all $k = 1, \ldots, n$. Then, in view of (5.27) and (5.33), the induced measure $\mathcal{Q}^n(x)$ can be expressed as

$$Z_N \mathcal{Q}^n(x) = \frac{\exp(-\beta N F^n(x))}{\prod_{k=1}^n (2\pi N)^{\frac{q-1}{2}} \sqrt{\det\left[\pi_k \operatorname{Hess} U_{|\Lambda_k|}(t^*(x^k/\pi_k))\right]_q}} (1 + o_N(1)). \tag{5.45}$$

Let us point out that the representation (5.45) of $\mathcal{Q}^n$ is in particular valid for all critical points of the free energy, F^n, since the essential smoothness of $I_{|\Lambda_k|}$, for all $k = 1, \ldots, n$, implies that these critical points are in the relative interior of $\bigtimes_{k=1}^n \mathcal{M}_{\pi_k} \equiv M$.

5.2.2. Free energy landscape near the critical point. In what follows, we will precisely compute the behavior of the measure $\mathcal{Q}^n$ in the neighborhood of critical points of F^n. First of all notice that the free energy is essential smooth and hence differentiable throughout $\mathrm{ri}(M)$. Thus, $x \in \mathrm{ri}(M)$ is a critical point if and only if $dF^n(x) = 0 \in T_x^* M$. In this particular case, the tangent space $T_x M$ coincides for all $x \in M$ with the subspace V_1 where we decompose $\mathbb{R}^{n \cdot q}$ into a direct sum of

$$V_0 := \mathrm{span}\{e^1 \otimes 1, \ldots, e^n \otimes 1\} \quad \text{and} \quad V_1 := V_0^\perp. \tag{5.46}$$

5.2. Induced measure and free energy landscape

Since M is embedded in the Euclidean space $\mathbb{R}^{n \cdot q}$, at a critical point $\boldsymbol{x} \in M$ there exists a $\boldsymbol{w} \equiv \boldsymbol{w}(\boldsymbol{x}) \in \mathbb{R}^{n \cdot q}$ such that $0 = \mathrm{d}F^n(\boldsymbol{x})(\boldsymbol{v}) = \langle \boldsymbol{w}, \boldsymbol{v} \rangle = \sum_{k=1}^n \langle w^k, v^k \rangle$ for all $\boldsymbol{v} \in \boldsymbol{V}_1$. This implies that $w^k \in \operatorname{span}\{1\} = V_0$. Recall that, for all k, the subdifferential of $I_{|\Lambda_k|}$ is a affine subspace of $\mathbb{R}^q$ with difference space V_0. Hence, a critical point is characterized by the solution of the system of equations

$$\operatorname{span}\{1\} \ni -2 \sum_{k=1}^n x^k - \bar{h}^k + \frac{1}{\beta} \partial I_{|\Lambda_k|}(x^k/\pi_k), \qquad \forall\, k = 1, \ldots, n \qquad (5.47)$$

Hence, for any $t^*(x^k/\pi_k) \in \partial I_{|\Lambda_k|}$ and $a \in \mathbb{R}$

$$\frac{1}{\beta} t^*(x^k/\pi_k) = 2 \sum_{k=1}^n x^k + \bar{h}^k + \gamma 1, \qquad \forall\, k = 1, \ldots, n. \qquad (5.48)$$

Making again use of the fact that $\nabla U_{|\Lambda_k|}(t^*(x^k/\pi_k)) = x^k/\pi_k$, (5.47) is equivalent to

$$x^k = \pi_k \nabla U_{|\Lambda_k|}(2\beta x + \beta\bar{h}^k), \qquad \forall\, k = 1, \ldots, n \qquad \text{with} \qquad x = \sum_{k=1}^n x^k, \qquad (5.49)$$

where we set for convenience $\gamma = 0$ since, by (5.28), the gradient $\nabla U_{|\Lambda_k|}$ is constant in direction $1 \in \mathbb{R}^q$. By summing over k, we see that solutions of the system of equations (5.49) are generated by the solutions of the equation

$$x = \sum_{k=1}^n \pi_k \nabla U_{|\Lambda_k|}(2\beta x + \beta\bar{h}^k) \qquad (5.50)$$

Let us denote a solution of (5.49) and (5.50) by $\boldsymbol{z}$ and z, respectively. The observation that a critical point $\boldsymbol{z}$ of F^n is already determined by the solution of the lower dimensional problem (5.50) is not a coincidence.

As we will show below, the structure of the free energy landscape, F^n, is closely related to a $q-1$-dimensional landscape which is given to leading order by the rate function, $F_N : \mathbb{R}^q \to \mathbb{R}$,

$$x \mapsto F_N(x) := -\|x\|^2 + \frac{1}{\beta} I_N(x) = -\|x\|^2 + \frac{1}{\beta} \sup_{t \in \mathbb{R}^q} \left(\langle t, x \rangle - U_N(t) \right)$$

of $\frac{1}{\beta N} \ln Q_N(x)$ where $U_N : \mathbb{R}^q \to \mathbb{R}$ is the log-moment generating function defined by

$$t \mapsto U_N(t) := \frac{1}{N} \ln \mathbf{E}^h_{\sigma[N]}\left[\exp(N \langle t, \varrho_N \rangle) \right] + \ln Z^h_N$$

$$= \frac{1}{N} \sum_{i=1}^N \ln \left(\sum_{r \in S_0} \tfrac{1}{q} \exp(\beta h^i_r + t_r) \right)$$

As a consequence of the definition of $U_{|\Lambda_k|}$, given in (5.17), and U_N, the equation (5.50) is equivalent to $x = \nabla U_N(2\beta x)$, i.e. it has the pleasant feature to be independent of the choice of the coarse graining. In the sequel we call F_N the *macroscopic free energy*.

In the following lemma, we collect some fact about the connection between F^n and F_N.

Lemma 5.8. *The functions F_N and F^n are related in the following ways.*

(i) *If $\mathbf{z}$ is a critical point of F^n, then $z \equiv z(\mathbf{z}) = \sum_{k=1}^n z^k$ is a critical point of F_N.*

(ii) *If z is a critical point of F_N, then $\mathbf{z} \equiv \mathbf{z}(z)$ where its components are given by $z^k = \pi_k \nabla U_{|\Lambda_k|}(2\beta z + \beta \bar{h}^k)$ is a critical point of F^n*

(iii) *At a critical point $\mathbf{z}$, it holds that*

$$F^n(\mathbf{z}) = F_N(z(\mathbf{z})) = \|z(\mathbf{z})\|^2 - \frac{1}{\beta N} \sum_{i=1}^N \ln\left(\sum_{r \in S_0} \tfrac{1}{q} \exp(2\beta z_r(\mathbf{z}) + \beta h_r^i)\right). \tag{5.51}$$

(iv) *For any $x \in \mathrm{ri}(\mathcal{M}_1)$,*

$$F_N(x) = \inf_{\substack{\mathbf{x} \in \times_{k=1}^n \mathcal{M}_{\pi_k}: \\ \sum_{k=1}^n x^k = x}} F^n(\mathbf{x}). \tag{5.52}$$

Proof. By a computation analog to the one presented in the lines (5.47) – (5.49), a critical point of F_N is determined by the solution of $x = \nabla U_N(2\beta x)$. Hence, (i) and (ii) are an immediate consequence of (5.49) and (5.50). In order to prove (iii), notice that (5.48) and (5.29) implies that

$$I_{|\Lambda_k|}(z^k/\pi_k) = \frac{\beta}{\pi_k} \langle 2z + \bar{h}^k, z^k \rangle - U_{|\Lambda_k|}(2\beta z + \beta \bar{h}^k), \qquad \forall\, k = 1, \ldots, n$$

where $z = \sum_{k=1}^n z^k$. By combining this expression with (5.15), we obtain that

$$\begin{aligned}
F^n(\mathbf{z}) &= -\|z\|^2 - \sum_{k=1}^n \langle \bar{h}^k, z^k \rangle + \sum_{k=1}^n \left(\langle 2z + \bar{h}^k, z^k \rangle - \frac{\pi_k}{\beta} U_{|\Lambda_k|}(2\beta z + \beta \bar{h}^k)\right) \\
&= \|z\|^2 - \frac{1}{\beta} \sum_{k=1}^n \pi_k\, U_{|\Lambda_k|}(2\beta z + \beta \bar{h}^k) \\
&= \|z\|^2 - \frac{1}{\beta N} \sum_{i=1}^N \ln\left(\sum_{r \in S_0} \tfrac{1}{q} \exp(2\beta z_r + \beta h_r^i)\right). \tag{5.53}
\end{aligned}$$

On the other hand, notice that $z = \nabla U_N(2\beta z)$ implies that $t^*(z) = 2\beta z + \gamma \mathbf{1}$. In view of (5.29), we have that $I_N(z) = 2\beta \|z\|^2 - U_N(2\beta z)$ which concludes the proof of (iii). Analog to the considerations leading to (5.49), for a given $x \in \mathrm{ri}(\mathcal{M}_1)$ the minimum on the set $\{\mathbf{x} \in \times_{k=1}^n \mathcal{M}_{\pi_k} \mid \sum_{k=1}^n x^k = x\}$ is attained at $\mathbf{x}(x)$ determined by the system of equations

$$x^k(x) = \pi_k \nabla U_{|\Lambda_k|}\left(\beta\left(2x + \bar{h}^k + \lambda \xi\right)\right), \qquad \forall\, k = 1, \ldots n, \tag{5.54}$$

5.2. Induced measure and free energy landscape

with $\xi := \frac{1}{\sqrt{q}}(q\,e^q - 1) \in \mathbb{R}^q$. By summing over all k, we obtain that the Lagrange multiplier, $\lambda \equiv \lambda(x)$, solves

$$x = \sum_{k=1}^{n} \pi_k \nabla U_{|\Lambda_k|}\left(\beta\left(2x + \bar{h}^k + \lambda\xi\right)\right) = \nabla U_N\left(\beta\left(2x + \lambda\xi\right)\right). \tag{5.55}$$

By convex duality (5.30), we have that $t^*\left(x^k(x)/\pi_k\right) = \beta\left(2x + \bar{h}^k + \lambda\xi\right)$ for all $k = 1, \ldots, n$ and $t^*(x) = \beta\left(2x + \lambda\xi\right)$. Thus, an computation analog to (5.51) reveals that

$$F^n(\boldsymbol{x}(x)) = -\|x\|^2 + \frac{1}{\beta}\left(\langle\beta(2x+\lambda\xi), x\rangle - U_N\left(\beta\left(2x+\lambda\xi\right)\right)\right) = F_N(x)$$

which concludes the proof. □

Remark 5.9. By the strong law of large numbers, the set of critical points of F_N converge, $\mathbb{P}_h$-a.s., to the set of solutions of the equation

$$z_r^* = \mathbb{E}\left[\frac{\exp(2\beta z_r^* + \beta h_r^1)}{\sum_{s \in \mathcal{S}_0} \exp(2\beta z_s^* + \beta h_s^1)}\right] \qquad \forall\, r \in \mathcal{S}_0, \tag{5.56}$$

i.e. in some sense the Curie-Weiss-Potts model in a random field is less disordered.

Remark 5.10. Let us consider a curve $\gamma \colon [0, 1] \to \mathcal{M}_1$ in the macroscopic free energy landscape that connects two minima and passes through the minimal saddle point between them. The corresponding minimal energy curve $\boldsymbol{\gamma}(t)$ defined by (5.54) and (5.55) has the property that it passes as well through the corresponding minima and the saddle point in the mesoscopic free energy landscape. Suppose there exists an $\varepsilon > 0$ such that the minimal component of $\gamma(t)$, for all $t \in [0, 1]$, is larger than ε. Then, the assumption on the random field $\{h^i\}$ implies that there exists two universal constants $0 < c \le C < \infty$ such that

$$c\,\pi_k\,\|\gamma(t)\| \;\le\; \left\|\frac{d}{dt}\boldsymbol{\gamma}(t)^k\right\| \;\le\; C\,\pi_k\,\|\gamma(t)\|, \tag{5.57}$$

uniformly in N and in $k = 1, \ldots, n$.

Next, we analyze the structure of the critical points. Mind that in a local coordinate system (5.30) reads

$$\mathbb{R}^{q-1} \ni s^* = \nabla\left(I_{|\Lambda_k|} \circ \psi^{-1}\right)(y) \quad\Longleftrightarrow\quad \mathbb{R}^{q-1} \ni y = \nabla\left(U_{|\Lambda_k|} \circ \psi^{-1}\right)(s^*).$$

Here, we choose as a chart $\psi \colon \mathcal{M}_1 \to \mathbb{R}^{q-1}$ the linear map $\psi(x) = V^\mathrm{T} x$ where the columns of the matrix V consist of the vectors of an orthonormal basis of V_1. Since the gradient map $\nabla U_{|\Lambda_k|}$ is invariant with respect to the subspace V_0, we restricted $U_{|\Lambda_k|}$ to $\mathrm{aff}(\mathcal{M}_1)$. Thus, we get that for any $x \in \mathrm{ri}(\mathcal{M}_1)$

$$\mathrm{Hess}\left(I_{|\Lambda_k|} \circ \psi^{-1}\right)(\psi(x)) = \left(V^\mathrm{T}\,\mathrm{Hess}\,U_{|\Lambda_k|}(t^*(x))\,V\right)^{-1}, \qquad \forall\, k = 1, \ldots, n. \tag{5.58}$$

Hence, the Hessian of the free energy, F^n, at a critical point $z \in \times_{k=1}^{n} \mathcal{M}_{\pi_k}$ is given in a local coordinate system, $\varphi(x) = I_n \otimes V^{\mathrm{T}}$, by

$$A^n(z) \equiv \mathrm{Hess}\left(F^n \circ \varphi^{-1}\right)(\varphi(z)) \;=\; 1 \cdot 1^{\mathrm{T}} \otimes (-2I_{q-1}) + \mathrm{diag}(U_1^{-1},\ldots,U_n^{-1}) \tag{5.59}$$

where $I_{q-1} \in \mathbb{R}^{q-1 \times q-1}$ denotes the identity matrix and

$$U_k \equiv U_k(z^k/\pi_k) := \beta\pi_k \, V^{\mathrm{T}} \, \mathrm{Hess}\, U_{|\Lambda_k|}(2\beta z + \beta \bar{h}^k) \, V. \tag{5.60}$$

We are interested in the behavior of $\mathcal{Q}^n$ in a neighborhood of the critical points of F^n. Suppose that z is a critical point of F^n. Then, for all $v \in V_1$ such that $\|v^k\| \le N^{-1/2+\delta}$ for all k

$$\frac{\mathcal{Q}^n(z+v)}{\mathcal{Q}^n(z)} \;=\; \exp\!\left(-\tfrac{\beta N}{2}\langle v, \mathcal{A}^n(z)\, v\rangle\right)\left(1 + o_N(1)\right) \tag{5.61}$$

provided that N large enough to ensure that $z^k + v^k \in \mathrm{ri}(\mathcal{M}_1)$. This is an immediate consequence of (5.45) and (5.48). Here, we set $\mathcal{A}^n(z) = (I_n \otimes V)\, A^n(z)\, (I_n \otimes V^{\mathrm{T}})$.

As a next step, we describe the eigenvalues of $A^n(z)$ at a critical point z of F^n. To start with, let us denote by $\{\xi_1, \ldots, \xi_K\}$ the distinct values that appear in the union of the spectrum of $U_k(z^k/\pi_k)^{-1}$. Since $z \in \mathrm{ri}(\times_{k=1}^{n} \mathcal{M}_{\pi_k})$ and in view of Lemma 5.6 all ξ_i's are non-zero. Further, let $\phi^{k,r}$ the eigenvector corresponding to the eigenvalue λ_r^k of $U_k(z^k/\pi_k)^{-1}$ and set $\kappa(\xi_l) := \{(k,r)\,|\,\lambda_r^k = \xi_l\}$.

Lemma 5.11. *Let z be a solution of the equation (5.50). Then γ is an eigenvalue of $A^n(z)$ if and only if either $\gamma = \xi_l$ for all $l \le K$ which satisfy*

$$|\kappa(\xi_l)| \;-\; \dim\left(\mathrm{span}\{\phi^{k,r}\,|\,(k,r) \in \kappa(\xi_l)\}\right) \;=:\; m_l \;>\; 0, \tag{5.62}$$

where m_l is the geometric multiplicity of ξ_l or γ is a solution of the equation

$$\det\!\left(I_{q-1} - 2\sum_{k=1}^{n}\left(U_k(z^k/\pi_k)^{-1} - \gamma\right)^{-1}\right) \;=\; 0. \tag{5.63}$$

Moreover, (5.63) has exactly one negative simple solution, i.e. γ is a simple root, if and only if, z is a critical point of index 1.

Proof. Let ξ_l be such that $m_l > 0$, if such a ξ_l exists. Then we will construct m_l orthogonal solutions of

$$\left(A^n(z) - \gamma\right)v \;=\; -2\left(1 \otimes \sum_{k=1}^{n} v^k\right) + \mathrm{diag}\!\left(U_1^{-1}-\gamma,\ldots,U_n^{-1}-\gamma\right) v \;=\; 0. \tag{5.64}$$

with eigenvalue $\gamma = \xi_l$. Namely, we set $0 \ne v = \sum_{(k,r)\in \kappa(\xi_l)} a_r^k (e^k \otimes \phi^{k,r})$ with $a_r^k \in \mathbb{R}$ which implies that (5.64) is equivalent to $\sum_{(k,r)\in \kappa(\xi_l)} a_r^k \phi^{k,r} = 0$. Notice that this equation has m_l orthogonal solutions. Hence, doing this for every ξ_l with $m_l > 0$, we can construct altogether $\sum_{l=1}^{k} m_l = n(q-1) - K$ eigenvectors of $A^n(z)$.

5.2. Induced measure and free energy landscape

Now, let us assume that γ is not an eigenvalue of U_k^{-1}, i.e. $\gamma \notin \{\xi_1, \ldots, \xi_K\}$. To find an eigenvalue of $A^n(z)$, an elementary computation shows that

$$0 = \det\left(A^n(z) - \gamma\right) = \prod_{k=1}^n \det\left(U_k^{-1} - \gamma\right) \cdot \det\left(I_{q-1} - 2 \sum_{k=1}^n \left(U_k^{-1} - \gamma\right)^{-1}\right), \tag{5.65}$$

Hence, (5.63) is then just the demand that the second term of (5.65) vanishes.

It remains to characterize the solutions of (5.63). Let $M(\gamma) = I_{q-1} - 2\sum_{k=1}^n (U_k^{-1} - \gamma)^{-1}$. Since the matrices U_k are positive definite, the matrix $M(\gamma)$ is regular if $\gamma \in (-\infty, 0]$. Clearly, $M(\gamma)$ converge to the identity matrix as γ tends to $-\infty$. On the other hand, for $\gamma_1, \gamma_2 \in (-\infty, 0]$ with $\gamma_1 < \gamma_2$ it holds that

$$M(\gamma_1) - M(\gamma_2) = 2(\gamma_2 - \gamma_1) \sum_{k=1}^n (U_k - \gamma_1)^{-1}(U_k - \gamma_2)^{-1} \tag{5.66}$$

i.e. the matrix $M(\gamma_1) - M(\gamma_2)$ is positive definite. In such a case, we also write $M(\gamma_1) \succ M(\gamma_2)$. Let us order the eigenvalues of $M(\gamma_i)$ in such a way that

$$\lambda_1(M(\gamma_i)) \leq \lambda_2(M(\gamma_i)) \leq \ldots \leq \lambda_{q-1}(M(\gamma_i)) \quad \text{for} \quad i=1,2.$$

By using the Theorem of Courant-Fischer, we obtain that $M(\gamma_1) \succ M(\gamma_2)$ implies that the eigenvalues satisfy $\lambda_r(M(\gamma_1)) > \lambda_r(M(\gamma_2))$ for all r. Together with the fact that the eigenvalues depends continuously on the entries, $\lambda_r(M(\gamma))$, seen as functions of γ, is strictly monotonic decreasing. Thus, $M(\gamma)$ has exactly one negative simple eigenvalue if and only if $M(0)$ has exactly one negative eigenvalue. But due to the Theorem of Ostrowski [95], $M(0)$ has exactly one negative eigenvalue if and only if z is a critical point of index 1. □

Combining the previous observations, we arrive at the following proposition.

Proposition 5.12. *Let $\{z^N\}_{N \in \mathbb{N}}$ be a sequence of critical points of F_N which converge to a critical point z^* of the deterministic landscape, i.e. z^* is the solution of (5.56). For each z^N, let $\mathbf{z}^N$ be the corresponding critical point of F^n given by (5.50) and $\hat{\mathbf{z}}^N \in \Gamma^n$ its lattice point approximation. Then, for all but finitely many values of N*

$$Z_N \, Q^n(\hat{\mathbf{z}}^N) = \frac{\exp(-\beta N \, F_N(z))\,(1 + o_N(1))}{\prod_{k=1}^n (2\pi N)^{\frac{q-1}{2}} \sqrt{\det\left[\pi_k \, \mathrm{Hess} \, U_{|\Lambda_k|}(2\beta \mathbf{z}^N + \beta \bar{h}^k)\right]_q}}. \tag{5.67}$$

In particular, we have that

$$\prod_{k=1}^n \det\left[\pi_k \, \mathrm{Hess} \, U_{|\Lambda_k|}(2\beta \mathbf{z}^N + \beta \bar{h}^k)\right]_q = \frac{\det\left(I_q - 2\beta \, \mathrm{Hess} \, U_N(2\beta z^N)\right)}{q^n \, \beta^{n \cdot (q-1)} \det\left(A^n(\mathbf{z}^N)\right)}, \tag{5.68}$$

Moreover, $\mathbb{P}_h$-a.s., for all but finitely many N it holds that

$$\det\left(I_q - 2\beta \operatorname{Hess} U_N\left(2\beta z^N\right)\right) = \det\left(I_q - 2\beta\left(D(z^N) - \mathbb{E}_h\left[u \cdot u^T\right]\right)\right)(1 + o_N(1)). \tag{5.69}$$

Recall that $D = \operatorname{diag}(z_1^N, \ldots, z_q^N)$ and $u \equiv u^1(2\beta z^N) \in \mathbb{R}^q$ is defined componentwise by (5.21).

Proof. By combining (5.45) with (5.48) and (5.51), (5.67) is immediate. Hence, it remains to show the expression for the prefactor (5.68). By an elementary computation, the determinant of the matrix $A^n(z)$ is of the following form

$$\det\left(A^n(z^N)\right) = \prod_{k=1}^n \det\left(U_k^{-1}\right) \cdot \det\left(I_{q-1} - 2\sum_{k=1}^n U_k\right)$$

$$= \frac{\det\left(I_q - 2\beta \operatorname{Hess} U_N\left(2\beta z^N\right)\right)}{\prod_{k=1}^n \det(U_k)}. \tag{5.70}$$

Let $W \in \mathbb{R}^{q-1 \times q-1}$ be the matrix consisting of the first $q-1$ rows of V. Then, (5.60) and Lemma 5.6 implies

$$\det(U_k) = \frac{\det(W^T U_k W)}{\det(W^T W)} = q\beta^{q-1} \det\left[\pi_k \operatorname{Hess} U_{|\Lambda_k|}(2\beta z^N + \beta \bar{h}^k)\right]_q \tag{5.71}$$

where we used that $\det(W^T W) = \det(I_{q-1} - \frac{1}{q} 1 \cdot 1^T) = 1 - \frac{q-1}{q} = \frac{1}{q}$. By combining (5.70) and (5.71), we obtain (5.68).

As a consequence of (5.50), $z^N = \nabla U_N(2\beta z^N)$. Hence,

$$\operatorname{Hess} U_N(2\beta z^N) = \operatorname{diag}(z_1^N, \ldots, z_r^N) - \frac{1}{N}\sum_{i=1}^N u^i(2\beta z^N) \cdot u^i(2\beta z^N)^T.$$

Moreover, the function $\mathbb{R}^q \ni t \mapsto \frac{1}{N}\sum_{i=1}^N u^i(t) \cdot u^i(t)^T \in \mathbb{R}^{q \times q}$ converge $\mathbb{P}_h$-a.s. to the deterministic function $\mathbb{E}\left[u(t) \cdot u(t)^T\right]$ uniformly on compact subsets of $\mathbb{R}^q$. Since the determinant is a continuous function, we conclude (5.69). $\square$

5.3. Upper bounds on capacities

In this section we derive upper bounds on capacities. For this purpose, we adapt the strategy that was originally presented in [6]. It relies on the idea to bound the microscopic capacity between two disjoint subsets $A, B \subset S_N$ in terms of the corresponding capacity of suitable sets in the coarse grained space Γ^n. Proceeding this way, we reduce the problem to the construction of a reasonable mesoscopic test function. First, we specify a function that is *almost harmonic* in a small neighborhood of the relevant saddle point. As a next step, we produce a good test function for the mesoscopic Dirichlet form from it.

5.3. Upper bounds on capacities

5.3.1. Mesoscopic capacity and partition of the mesoscopic state space.
Let us consider two disjoint subsets $A, B \subset \Gamma^n$ and set $\boldsymbol{A} = \mathcal{S}^n[A]$ and $\boldsymbol{B} = \mathcal{S}^n[B]$. As an immediate consequence of the *Dirichlet principle*, we can bound the *microscopic capacity* $\mathrm{cap}(A, B)$ from above by the *mesoscopic capacity* $\mathrm{CAP}^n(\boldsymbol{A}, \boldsymbol{B})$

$$
\begin{aligned}
\mathrm{cap}(A, B) &= \inf_{h \in \mathcal{H}_{A,B}} \frac{1}{2} \sum_{\sigma, \eta \in \mathcal{S}_N} \mu_N(\sigma) p_N(\sigma, \eta) \left(h(\sigma) - h(\eta)\right)^2 \\
&\leq \inf_{g \in \mathcal{G}_{A,B}} \frac{1}{2} \sum_{\sigma, \eta \in \mathcal{S}_N} \mu_N(\sigma) p_N(\sigma, \sigma') \left(g(\varrho^n(\sigma)) - g(\varrho^n(\eta))\right)^2 \\
&= \inf_{g \in \mathcal{G}_{A,B}} \frac{1}{2} \sum_{x, y \in \Gamma^n} \mathcal{Q}^n(x) r^n(x, y) \left(g(x) - g(y)\right)^2 \\
&=: \mathrm{CAP}^n(\boldsymbol{A}, \boldsymbol{B}),
\end{aligned}
\qquad (5.72)
$$

where

$$
\mathcal{H}_{A,B} := \{h \colon \mathcal{S}_N \to [0, 1] \mid h|_A \equiv 1,\; h|_B \equiv 0\}, \qquad (5.73)
$$

$$
\mathcal{G}_{A,B} := \{g \colon \Gamma^n \to [0, 1] \mid g|_{\boldsymbol{A}} \equiv 1,\; g|_{\boldsymbol{B}} \equiv 0\}. \qquad (5.74)
$$

Now, let $\boldsymbol{z}$ be a saddle point of F^n with index 1, i.e. $\boldsymbol{z}$ is a critical point of F^n and $\mathbb{A}^n(\boldsymbol{z})$ has exactly one negative eigenvalue while all other eigenvalues are strictly positive. Further let $\boldsymbol{A}, \boldsymbol{B} \subset \Gamma^n$ be two subsets that are strictly contained in two different connected components of the level sets $\{\boldsymbol{x} \in \Gamma^n \mid F^n(\boldsymbol{x}) < F^n(\boldsymbol{z})\}$. Additionally, $\boldsymbol{A}, \boldsymbol{B}$ are chosen in such a way that there exists a path γ from $\boldsymbol{A}$ to $\boldsymbol{B}$ such that $\max_{\boldsymbol{x} \in \gamma} F^n(\boldsymbol{x}) = F^n(\boldsymbol{z})$.

Let us point out that only those points in $\boldsymbol{x}, \boldsymbol{y} \in \Gamma^n$ contribute to the capacity for which $r^n(\boldsymbol{x}, \boldsymbol{y}) > 0$, the modulus of the difference of the harmonic function between $\boldsymbol{x}, \boldsymbol{y}$ is large and the induced measure, $\mathcal{Q}^n(\boldsymbol{x})$, is not to small. In [15] it was shown that the harmonic function typically changes from one to zero in a small neighborhood of the saddle point $\boldsymbol{z}$, see Figure 5.3.1. Hence, for $\delta_N = c_0\, N^{-1/2+\delta}$ with $c_0 < \infty$ and $\delta > 0$ let us consider the domain

$$
D^n \equiv D^n(\boldsymbol{z}, \delta_N) := \{\boldsymbol{x} \in \Gamma^n \mid \|\boldsymbol{z} - \boldsymbol{x}\|_\infty \leq \delta_N\}. \qquad (5.75)
$$

Here, $D^n(\boldsymbol{z}, \delta_N)$ is a cube in Γ^n centered in $\boldsymbol{z}$ with side length $2\delta_N$. Further, for a fixed vector $\boldsymbol{v} \in \Gamma^n$, to be defined below, consider the subsets

$$
\begin{aligned}
W_0 &:= \{\boldsymbol{x} \in \Gamma^n \mid |\langle \boldsymbol{v}, \boldsymbol{x} - \boldsymbol{z}\rangle| < \delta_N \} \\
W_1 &:= \{\boldsymbol{x} \in \Gamma^n \mid \langle \boldsymbol{v}, \boldsymbol{x} - \boldsymbol{z}\rangle \leq -\delta_N\}, \\
W_2 &:= \{\boldsymbol{x} \in \Gamma^n \mid \langle \boldsymbol{v}, \boldsymbol{x} - \boldsymbol{z}\rangle \geq \delta_N \},
\end{aligned}
\qquad (5.76)
$$

Our strategy is to compute first an upper bound on the capacity restricted to the set D^n using an appropriate test function that is almost harmonic in D^n with boundary

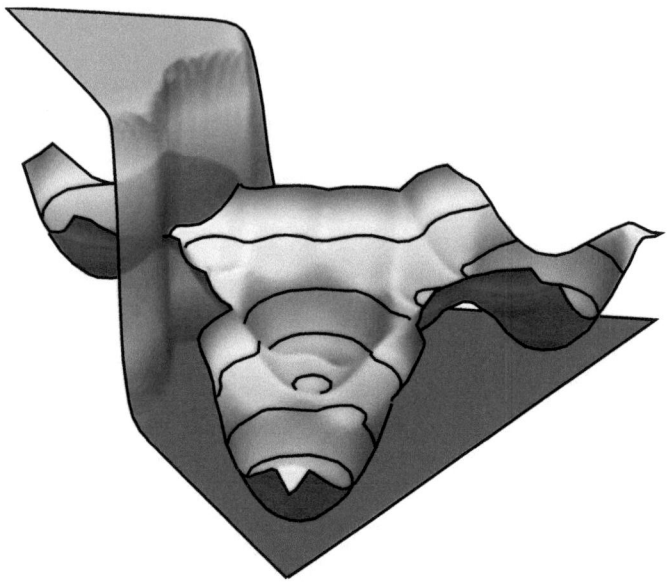

Figure 1. Comparison of the free energy landscape and the harmonic function in the random field Curie-Weiss Potts model for $(q = 3)$ with boundary condition 1 on the shallow local minimum and 0 on the global minimum.

condition zero and one, respectively, on the sets $W_1 \cap D^n$ and $W_2 \cap D^n$. Afterwards, we will show that the contribution to the capacity outside D^n is negligible.

5.3.2. Distortion by mesoscopic transition probabilities. In order to construct a test function that is almost harmonic inside D^n, we have to take into account the Hessian of the free energy, F^n, at the saddle point, z, as well as the mesoscopic transition probabilities. In what following, we first adopted the construction of a distortion of the form $B^n(z) = G^T A^n(z) G$ that was originally introduced in [6]. In a second step, we characterize the eigenvalues of $B^n(z)$ and some aspects of the corresponding eigenvectors.

We begin with considering the block diagonal matrix $R = \text{diag}(R_1, \ldots, R_n) \in \mathbb{R}^{n \cdot q \times n \cdot q}$ where the matrices $R_k \in \mathbb{R}^{q \times q}$, for each $k = 1, \ldots, n$, are given by

$$R_k \equiv R_k(z) := \sum_{(r,s) \in \Delta_k} r^n(z, z - \hat{e}^{k,r} + \hat{e}^{k,s}) (e^r - e^s) \cdot (e^r - e^s)^T. \quad (5.77)$$

Here, z is the saddle point, specified above, and $\Delta_k \subset \mathcal{S}_0^2$, is the smallest set containing all undirected edges of $\mathcal{S}_0^2$, e.g. $\Delta_k = \{(r,s) \in \mathcal{S}_0^2 \mid s < r\}$. We denote by $\hat{e}^{k,r} \equiv \frac{1}{N} e^k \otimes e^r$ a coordinate vector in $\mathbb{R}^{n \cdot q}$ of length $\frac{1}{N}$. A characterization of Δ_k for $k = 1, \ldots, n$

5.3. Upper bounds on capacities

will be given in Assumption 5.18. Obviously, for each k, the matrix R_k is positive semi-definite and has an eigenvalue zero corresponding to the eigenvector $1 \in \mathbb{R}^q$.

Before we characterize the remaining eigenvalues of R_k, let us have a closer look at the mesoscopic transition probabilities. As a consequence of the underlying microscopic single-site dynamics, for any $x, y \in \Gamma^n$ only those transitions $r^n(x, y)$ may be different from zero, for which $y = x - \hat{e}^{k,r} + \hat{e}^{k,s}$ with $k \in \{1, \ldots, n\}$ and $r, s \in \mathcal{S}_0$. For a given $\sigma \in \mathcal{S}_N$, let us denote by

$$\Lambda_{k,r}(\sigma) := \{i \in \Lambda_k \,|\, \sigma_i = r\} \tag{5.78}$$

the set of lattice sites i in the block Λ_k those spin variable σ_i are equal to the color r. For $x \in \Gamma^n$ set $x = \sum_{k=1}^{n} x^k$. Then, we have

$$r^n\left(x, x - \hat{e}^{k,r} + \hat{e}^{k,s}\right)$$

$$= \frac{1}{Q^n(x)} \sum_{\sigma \in \mathcal{S}^n[x]} \mu_N(\sigma) \sum_{i \in \Lambda_{k,r}(\sigma)} p_N(\sigma, \sigma^{i,s})$$

$$= \frac{1}{Q^n(x)} \sum_{\sigma \in \mathcal{S}^n[x]} \mu_N(\sigma) \sum_{i \in \Lambda_{k,r}(\sigma)} \frac{\exp\left(-\beta \left[\langle 2x + h^i, e^r - e^s\rangle - \frac{2}{N}\right]_+\right)}{Nq},$$

where the configuration $\sigma^{i,r}$ is obtained from σ by replacing the color σ_i at site i through the color $r \in \mathcal{S}_0$. Notice that for all $x \in \Gamma^n$ and $\sigma \in \mathcal{S}^n[x]$ the cardinality of the set $\Lambda_{k,r}(\sigma)$ is equal to Nx_r^k. In particular, $|\Lambda_{k,r}(\sigma)|$ assumes the same value for all $\sigma \in \mathcal{S}^n[x]$. Using that $h^i = \bar{h}^k + \tilde{h}^i$ and $\operatorname{diam} \mathcal{H}_k \leq \varepsilon(n)$, we get for $i \in \Lambda_k$ the estimate

$$r^n\left(x, x - \hat{e}^{k,r} + \hat{e}^{k,s}\right) = \frac{x_r^k}{q} \exp\left(-\beta \left[\langle 2x + \bar{h}^k, e^r - e^s\rangle - \frac{2}{N}\right]_+\right) (1 + \mathcal{O}(\varepsilon(n))). \tag{5.79}$$

Thus, provided N is large enough, (5.79) implies that there exists a $c_{15}(n) > 0$ independent of N such that the mesoscopic rates at the saddle point z satisfy $r^n(z, z - \hat{e}^{k,r} + \hat{e}^{k,s}) \geq c_{15}(n)$ for all $k = 1, \ldots, n$ and $r, s \in \mathcal{S}_0$.

Let us know come to a characterization of the eigenvalues in the orthogonal complement, V_1, to the subspace $\operatorname{span}\{1\}$.

Lemma 5.13. *For each k, the eigenvalues of the matrix R_k with respect to the subspace V_1 are bounded from below by $c_{15}(n)$.*

Proof. The matrix R_k is symmetric. Hence, by exploiting the interlacing property for symmetric matrices, we can bound the eigenvalues of R_k with respect to V_1 by the eigenvalues of the leading principle minor, i.e. $\lambda_s(R_k) \leq \lambda_s([R_k]_q) \leq \lambda_{s+1}(R_k)$ for all $s = 1, \ldots, q-1$ whereas $\lambda_1(R_k) = 0$. By applying Gershgorin's Theorem and computations similar to the ones given in the proof of Lemma 5.6, the eigenvalues of the principle minor $[R_k]_q$ are bounded from below by

$$\lambda_s([R_k]_q) \geq r^n(z, z - \hat{e}^{k,q} + \hat{e}^{k,s})\,\mathbb{1}_{(q,s)\in\Delta_k} + r^n(z, z - \hat{e}^{k,s} + \hat{e}^{k,q})\,\mathbb{1}_{(s,q)\in\Delta_k}$$
$$\geq c_{15}(n), \qquad (5.80)$$

for all $s = 1, \ldots, q-1$. This completes the proof. □

As an immediate consequence of the lemma above we have that, for each k, the restriction of R_k to V_1 is regular. Moreover, there exists a positive semi-definite matrix G_k such that $R_k = G_k G_k^T$. The distortion of $A^n(z)$ by the mesoscopic transition probabilities r^n is now defined through

$$B^n(z) := \mathrm{diag}(G_1^T, \ldots, G_n^T)\, A^n(z)\, \mathrm{diag}(G_1, \ldots, G_n) = G^T A^n(z)\, G. \qquad (5.81)$$

Let us denote by $\mathcal{B}^n(z) = (I_n \otimes V^T)\, B^n(z)(I_n \otimes V)$ and $\mathcal{G} = (I_n \otimes V^T)\, G\, (I_n \otimes V)$ the corresponding projections to $\boldsymbol{V}_1$.

Remark 5.14. In comparison with the distortion defined in [6], it turns out that the block diagonal form of the matrix G reflects the coarse graining procedure whereas the particular form of the matrix R_k is linked to the underlying graph structure of the mesoscopic transition probabilities.

Let $\{\hat{v}^1, \ldots, \hat{v}^{n\cdot(q-1)}\}$ be an orthonormal basis of eigenvectors to the eigenvalues $\hat{\gamma}_k$ of $\mathcal{B}^n(z)$ with respect to $\boldsymbol{V}_1$. Further define the distorted vectors

$$v^k := \mathcal{G}^{-T} \hat{v}^k \quad \text{and} \quad \check{v}^k := \mathcal{G} \hat{v}^k = \mathcal{R} v^k. \qquad (5.82)$$

Here, $\mathcal{G}^{-1}$ denotes the Moore-Penrose pseudo inverse of the matrix $\mathcal{G}$, i.e. it is defined by the property that $\mathcal{G}\mathcal{G}^{-1}\mathcal{G} = \mathcal{G}$, $\mathcal{G}^{-1}\mathcal{G}\mathcal{G}^{-1} = \mathcal{G}^{-1}$ and $\mathcal{G}\mathcal{G}^{-1}$ is symmetric.

Remark 5.15. Notice that the matrix $\mathcal{G}$ restricted to the subspace $\boldsymbol{V}_1$ is regular, and hence the Moore-Penrose pseudo inverse does not cause any further difficulties.

An important fact about these vectors is that

$$\langle \check{v}^l, v^k \rangle = \delta(l,k) \quad \text{and} \quad \mathcal{A}^n(z)\, \check{v}^k = \hat{\gamma}_k\, v^k. \qquad (5.83)$$

This implies the following non-orthogonal decomposition of the quadratic form

$$\langle x, \mathcal{A}^n(z)\, y \rangle = \sum_{k=1}^{n\cdot(q-1)} \hat{\gamma}_k \langle x, v^k \rangle \langle v^k, y \rangle, \qquad x, y \in \boldsymbol{V}_1. \qquad (5.84)$$

In the following lemma we characterize the eigenvalues of $\mathcal{B}^n(z)$.

5.3. Upper bounds on capacities

Lemma 5.16. *Let z be a solution of the equation (5.50) and z be given by (5.49). Further, assume that z is a critical point of index 1. Then, $B^n(z)$ has a unique negative eigenvalue $\hat{\gamma}_1 \equiv \hat{\gamma}_1(N, n)$ which is given by the negative solution of the equation*

$$\det \left(I_{q-1} - 2 \sum_{k=1}^{n} \left(U_k(z^k/\pi_k)^{-1} - \gamma R_k^{-1} \right)^{-1} \right) = 0. \quad (5.85)$$

Moreover, we have that, $\mathbb{P}_h$-a.s,

$$\lim_{n \to \infty} \lim_{N \to \infty} \hat{\gamma}_1(N, n) =: \bar{\gamma}_1, \quad (5.86)$$

where $\bar{\gamma}_1$ is the unique negative solution of the equation

$$\det \left(I_{q-1} - 2\beta \, \mathbb{E} \left[\left((V^{\mathrm{T}}(\mathrm{diag}(u_1, \ldots, u_q) - u \cdot u^{\mathrm{T}})V \right)^{-1} - \gamma \left(V^{\mathrm{T}} \bar{R} V \right)^{-1} \right)^{-1} \right] \right) = 0, \quad (5.87)$$

with $u \equiv u^1(2\beta z)$, whereas the function u^i is defined in (5.21) and

$$\bar{R} = \sum_{(r,s) \in \Delta_k} \frac{\exp(2\beta z_r + \beta h_r^1) \exp(-\beta [\langle 2z + h^1, e^r - e^s \rangle]_+)}{\beta q \sum_{r' \in S_0} \exp(2\beta z_{r'} + \beta h_{r'}^1)} (e^r - e^s) \cdot (e^r - e^s)^{\mathrm{T}}. \quad (5.88)$$

Proof. Since the matrix G is positive definite, the Theorem of Ostrowski implies that there exist positive numbers θ_i such that $\lambda_i(B^n(z)) = \theta_i \lambda_i(A^n(z))$. By combining this fact with Lemma 5.11, we deduce that if z is a critical point of index 1, $B^n(z)$ has a unique negative eigenvalue. Further set $R_k = G_k G_k^{\mathrm{T}}$. Then, for $\gamma \in (\infty, 0]$,

$$0 = \det \left(B^n(z) - \gamma \right)$$

$$= \prod_{k=1}^{n} \det \left(R_k U_k^{-1} - \gamma \right) \cdot \det \left(I_{q-1} - 2 \sum_{k=1}^{n} \left(R_k U_k^{-1} - \gamma \right)^{-1} R_k \right).$$

The fact that the eigenvalues of $R_k U_k^{-1}$ are positive implies that the unique negative eigenvalue $\hat{\gamma}_1$ annihilate the last determinant. Hence, we recover (5.85).

It remains to show the convergence property. Inserting the expression for U_k, given by (5.60), reveals that (5.85) is equivalent to the fact that the determinant

$$\det \left(I_{q-1} - 2\beta \sum_{k=1}^{n} \pi_k \left((V^{\mathrm{T}} \mathrm{Hess} \, U_{|\Lambda_k|}(2\beta z + \beta \bar{h}^k) V)^{-1} - \gamma (V^{\mathrm{T}} \tfrac{1}{\beta \pi_k} R_k V)^{-1} \right)^{-1} \right)$$

vanishes. By substituting (5.49) into (5.79) we get

$$\frac{\frac{1}{\beta \pi_k} R_k}{(1 + \mathcal{O}(\varepsilon))} = \sum_{(r,s) \in \Delta_k} \frac{\exp(2\beta z_r + \beta \bar{h}_r^k) \exp(-\beta [\langle 2z + \bar{h}^k, e^r - e^s \rangle - \frac{2}{N}]_+)}{\beta q \sum_{r' \in S_0} \exp(2\beta z_{r'} + \beta \bar{h}_{r'}^k)} (e^r - e^s) \cdot (e^r - e^s)^{\mathrm{T}}. \quad (5.89)$$

On the other hand, a simple computation shows that

$$\frac{\operatorname{Hess} U_{|\Lambda_k|}(2\beta z + \beta \bar{h}^k)_{r,s}}{(1+\mathcal{O}(\varepsilon))}$$

$$= \left(\frac{\exp(2\beta z_r + \beta \bar{h}_r^k)}{\sum_{r' \in S_0} \exp(2\beta z_{r'} + \beta \bar{h}_{r'}^k)} \delta(r,s) - \frac{\exp(2\beta z_r + \beta \bar{h}_r^k)\exp(2\beta z_s + \beta \bar{h}_s^k)}{\left(\sum_{r' \in S_0} \exp(2\beta z_{r'} + \beta \bar{h}_{r'}^k)\right)^2} \right).$$

By exploiting the continuity of the determinant allows to deduce the claimed convergence property. □

A consequence of the particular form of the matrix $A^n(z)$, see (5.59), is the following

Lemma 5.17. *Let $v \equiv v^1$. Then, there exists a constant $c_{16} > 0$ such that, independent of n.*

$$c_{16} \leq \min_k \|v^k\| = \max_k \|v^k\| \leq \frac{1}{c_{16}}. \tag{5.90}$$

Proof. First of all notice that $B^n(z)\hat{v}^1 = \hat{\gamma}_1 \hat{v}^1$ is equivalent to $A^n(z)Rv^1 = \hat{\gamma}_1 v^1$. By (5.59) it follows that

$$2\sum_{l=1}^n R_l V^T v^l = \left(U_k^{-1} R_k - \hat{\gamma}_1\right) V^T v^k, \quad \forall k = 1,\ldots,n. \tag{5.91}$$

Recall that $V \in \mathbb{R}^{q \times (q-1)}$ is the matrix those columns consist of an orthonormal basis of V_1. As an immediate consequence of (5.91) we have that for all $k, l = 1, \ldots, n$

$$C_k v^k := \left(U_k^{-1} R_k - \hat{\gamma}_1\right) V^T v^k = \left(U_l^{-1} R_l - \hat{\gamma}_1\right) V^T v^l = C_l v^l \tag{5.92}$$

Hence, $\hat{v}^l = G_l^T C_l^{-1} C_k v^k$. On the other hand, $U_k^{-1} R_k = (V^T \operatorname{Hess} U_{|\Lambda_k|}(2\beta z + \beta \bar{h}^k) V)^{-1} \frac{1}{\beta \pi_k} R_k$. In view of Lemma 5.6 and (5.89) combined with Lemma 5.13 we conclude that the eigenvalues of the matrix C_k are bounded away from zero and infinity, uniformly in n, N and $k = 1, \ldots, n$. Moreover,

$$1 = \langle \hat{v}, \hat{v} \rangle = \sum_{l=1}^n \langle \hat{v}^l, \hat{v}^l \rangle = \left\langle v^k, C_k^T \left(\sum_{l=1}^n C_l^{-T} R_l C_l\right) C_k v^k \right\rangle. \tag{5.93}$$

By exploiting again the Theorem of Ostrowski, the assertion of the lemma follows. □

Concerning the construction of the matrices R_k there is some freedom in choosing the sets Δ_k. Let us point out that the eigenvalues of the matrix $B^n(z)$ as well as the corresponding eigenvectors may depend crucially on the corresponding choice. Further, let us emphasis that the computation of an upper bound on capacities is not sensitive to the particular choice of Δ_k, except for the fact that the unique negative eigenvalue $\hat{\gamma}_1$ appears in the resulting expression. However, in order to construct a non-negative unit flow that leads to a matching lower bound, the choice of Δ_k really

5.3. Upper bounds on capacities

matters. It turns out that we should construct the matrices R_k in such a way that the vector $v^1 \equiv v = \sum_{k=1}^{n} e^k \otimes v^k$ defined in terms of the eigenvector $\check{v}^1$ corresponding to the unique negative eigenvalue of the matrix $B^n(z)$ satisfies the property that

$$\text{either } v_r^k - v_s^k \geq 0 \quad \text{or} \quad v_r^k - v_s^k \leq 0 \qquad \forall k = 1, \ldots, n \text{ and } (r,s) \in \Delta_k. \tag{5.94}$$

One may expect that if we choose the sets Δ_k in such a way that the eigenvector to the negative eigenvalue of $A^n(z)$ satisfies (5.94), this property is preserved for the distorted vector v. However, a proof of this claim is challenging and missing so far. For this reason, we will rely in the sequel on the following

Assumption 5.18. *Assume that there exists sets $\Delta_1, \ldots, \Delta_n$ such that the vector v^1 that corresponds to the negative eigenvalue, $\hat{\gamma}_1$, of the matrix $B^n(z)$ satisfy the property*

$$v_r^k - v_s^k \geq 0 \qquad \forall k = 1, \ldots, n \text{ and } (r,s) \in \Delta_k. \tag{5.95}$$

5.3.3. Approximation of the harmonic function near the saddle point.
Now, our aim is to construct a function $g : \mathbb{R}^{n \cdot q} \to [0,1]$ which is almost harmonic with respect to the generator L^n. To start with, consider the function $f : \mathbb{R} \to [0,1]$ given by

$$f(s) = \sqrt{\frac{\beta N |\hat{\gamma}_1|}{2\pi}} \int_{-\infty}^{s} \exp\left(-\frac{\beta N |\hat{\gamma}_1|}{2} t^2\right) dt. \tag{5.96}$$

Then, we define the function g through

$$g(\boldsymbol{x}) = f(\langle v, \boldsymbol{x} - \boldsymbol{z}\rangle), \tag{5.97}$$

with $v \equiv v^1$ as defined in (5.82). Let us point out that this vector v is also used in the definition of the sets W_0, W_1, W_2. Due to the choice of δ_N, we have that in the limit when $N \to \infty$ the function $g(\boldsymbol{x})$ converge exponential fast to 0 for all $\boldsymbol{x} \in W_1 \cap \partial D^n(\boldsymbol{z}, \delta_N)$ and to 1 for all $\boldsymbol{x} \in W_2 \cap \partial D^n(\boldsymbol{z}, \delta_N)$.

For our further computations, the transition rates $r^n(\boldsymbol{x}, \boldsymbol{y})$ are in a slightly unpleasant form. Moreover, we would like to replace the measure $\mathcal{Q}^n$ in the neighborhood D^n of the saddle point z under consideration by the approximation given in (5.61). Based on the fact that the Dirichlet form is monotone in the transition probabilities, in [6] the following comparison result was established

Lemma 5.19 ([6, Lemma 4.1]). *Let $(r(x,y))_{x,y \in \Gamma}$ and $(\bar{r}(x,y))_{x,y \in \Gamma}$ be two transition matrices that are reversible with respect to the invariant distributions Q and $\bar{Q}$, respectively. Assume that for all $x, y \in \Gamma$, there exists $\delta > 0$ such that*

$$\left|\frac{Q(x)}{\bar{Q}(x)} - 1\right| \leq \delta \quad \text{and} \quad \left|\frac{r(x,y)}{\bar{r}(x,y)} - 1\right| \leq \delta. \tag{5.98}$$

Then, for any disjoint subsets $A, B \subset \Gamma$,

$$(1-\delta)^2 \leq \frac{\text{CAP}(A,B)}{\overline{\text{CAP}}(A,B)} \leq (1-\delta)^{-2}. \tag{5.99}$$

Hence, we are left with finding suitable modifications of the transition probabilities, $r^n(x,y)$, and the measure, $\mathcal{Q}^n(x)$, respectively. First, let us define the measure $\overline{\mathcal{Q}}^n$ through

$$\overline{\mathcal{Q}}^n(x) := \mathcal{Q}^n(z) \exp\left(-\tfrac{\beta N}{2} \langle x-z, A^n(z)(x-z)\rangle\right), \qquad x \in \Gamma^n.$$

In view of (5.61), there exists a $0 \leq c_{17} < \infty$ such that for all $x \in D^n$,

$$\left| \frac{\mathcal{Q}^n(x)}{\overline{\mathcal{Q}}^n(x)} - 1 \right| \leq c_{17} N \delta_N^3.$$

By (5.79), the mesoscopic transition probabilities satisfy for all $x \in D^n$ the bound

$$\frac{x_r^k}{z_r^k} e^{-4\beta\delta_N} e^{-2\beta\varepsilon(n)} \leq \frac{r^n(x, x - \hat{e}^{k,r} + \hat{e}^{k,s})}{r^n(z, z - \hat{e}^{k,r} + \hat{e}^{k,s})} \leq \frac{x_r^k}{z_r^k} e^{4\beta\delta_N} e^{2\beta\varepsilon(n)}.$$

By exploiting the fact that critical points are bounded away from the boundary of $\times_{k=1}^n \mathcal{M}_{\pi_k}$ implies that there exists $0 \leq c_{18} < \infty$ such that for all $x \in D^n$,

$$\left| \frac{r^n(x, x - \hat{e}^{k,r} + \hat{e}^{k,s})}{r^n(z, z - \hat{e}^{k,r} + \hat{e}^{k,s})} - 1 \right| \leq c_{18} \max\{\delta_N, \varepsilon(n)\}. \tag{5.100}$$

Hence, we define the following mesoscopic transition rates

$$\bar{r}^n(x, x - \hat{e}^{k,r} + \hat{e}^{k,s}) := r_{r,s}^k \quad \text{and} \quad \bar{r}^n(x - \hat{e}^{k,r} + \hat{e}^{k,s}, x) := \frac{r_{r,s}^k \overline{\mathcal{Q}}^n(x)}{\overline{\mathcal{Q}}^n(x - \hat{e}^{k,r} + \hat{e}^{k,s})}$$

with $r_{r,s}^k \equiv r^n(z, z - \hat{e}^{k,r} + \hat{e}^{k,s})$ having the property that they are reversible with respect to the measure $\overline{\mathcal{Q}}^n$. Moreover, we define the operator $\overline{L}^n$ acting on functions $g: \Gamma^n \to \mathbb{R}$ as

$$(\overline{L}^n g)(x) = \sum_{k=1}^n \sum_{r,s=1}^q \bar{r}^n(x, x - \hat{e}^{k,r} + \hat{e}^{k,s}) \left(g(x - \hat{e}^{k,r} + \hat{e}^{k,s}) - g(x) \right). \tag{5.101}$$

The actual advantage of choosing the negative eigenvalue, $\hat{\gamma}_1$, of $B^n(z)$ and the distorted eigenvector v^1 is that we can derive an estimate on $|(\overline{L}^n g)(x)|$ that is by a factor δ_N^2 smaller than an arbitrary choice of the parameters $\hat{\gamma}_1$ and v. We prove this fact in the following lemma.

Lemma 5.20. *Consider the function g as defined in (5.97). Then, for all $x \in D^n$, there exists a constant $0 \leq c < \infty$ such that*

$$|(\overline{L}^n g)(x)| \leq \sqrt{\tfrac{\beta|\hat{\gamma}_1|}{2\pi N}} \exp\left(-\tfrac{\beta N|\hat{\gamma}_1|}{2} \langle v, x-z\rangle^2\right) c \delta_N^2. \tag{5.102}$$

5.3. Upper bounds on capacities

Proof. To lighten the notation, we assume throughout this proof that the origin of the coordinate system is chosen in such a way that $z = 0$. Further, we set $A^n(z) \equiv A$. Now, for any $k = 1, \ldots, n$ and $(r,s) \in \Delta_k$, let us rewrite the corresponding terms in the generator (5.101) by means of the detailed balance condition as

$$r_{r,s}^k \left(g(\boldsymbol{x} - \hat{\boldsymbol{e}}^{k,r} + \hat{\boldsymbol{e}}^{k,s}) - g(\boldsymbol{x}) \right)$$

$$\times \left(1 + \frac{\overline{Q}^n(\boldsymbol{x} + \hat{\boldsymbol{e}}^{k,r} - \hat{\boldsymbol{e}}^{k,s})}{\overline{Q}^n(\boldsymbol{x})} \cdot \frac{g(\boldsymbol{x} + \hat{\boldsymbol{e}}^{k,r} - \hat{\boldsymbol{e}}^{k,s}) - g(\boldsymbol{x})}{g(\boldsymbol{x} - \hat{\boldsymbol{e}}^{k,r} + \hat{\boldsymbol{e}}^{k,s}) - g(\boldsymbol{x})} \right). \quad (5.103)$$

Since $\boldsymbol{x} \in D^n(0, \delta_N)$, we have that

$$\frac{\overline{Q}^n(\boldsymbol{x} + \hat{\boldsymbol{e}}^{k,r} - \hat{\boldsymbol{e}}^{k,s})}{\overline{Q}^n(\boldsymbol{x})}$$

$$= \exp\left(-\beta \langle \boldsymbol{e}^{k,r} - \boldsymbol{e}^{k,s}, A\boldsymbol{x} \rangle\right) \exp\left(-\tfrac{\beta}{2N} \langle \boldsymbol{e}^{k,r} - \boldsymbol{e}^{k,s}, A(\boldsymbol{e}^{k,r} - \boldsymbol{e}^{k,s}) \rangle\right)$$

$$= \exp\left(-\beta \langle \boldsymbol{e}^{k,r} - \boldsymbol{e}^{k,s}, A\boldsymbol{x} \rangle\right) (1 + \mathcal{O}(1/N)). \quad (5.104)$$

In view of (5.97), let us first consider the difference $f(a + \tfrac{1}{N}b) - f(a)$. A straight forward computation yields

$$f(a + \tfrac{1}{N}b) - f(a)$$

$$= \sqrt{\tfrac{\beta|\hat{\gamma}_1|}{2\pi N}} \exp\left(-\tfrac{\beta N |\hat{\gamma}_1|}{2} a^2\right) \int_0^b \exp\left(-\tfrac{\beta|\hat{\gamma}_1|}{2N} t^2 - \beta|\hat{\gamma}_1| a t\right) dt$$

$$= b \sqrt{\tfrac{\beta|\hat{\gamma}_1|}{2\pi N}} \exp\left(-\tfrac{\beta N |\hat{\gamma}_1|}{2} a^2\right) \exp\left(-\tfrac{\beta|\hat{\gamma}_1|}{2} ab\right) \frac{\sinh\left(\tfrac{\beta|\hat{\gamma}_1|}{2} ab\right)}{\tfrac{\beta|\hat{\gamma}_1|}{2} ab} (1 + \mathcal{O}(\tfrac{1}{N})).$$

Notice that the function $\sinh(x)/x$ is symmetric. Hence,

$$\frac{g(\boldsymbol{x} - \hat{\boldsymbol{e}}^{k,r} + \hat{\boldsymbol{e}}^{k,s}) - g(\boldsymbol{x})}{g(\boldsymbol{x} + \hat{\boldsymbol{e}}^{k,r} - \hat{\boldsymbol{e}}^{k,s}) - g(\boldsymbol{x})} = -\exp\left(\beta|\hat{\gamma}_1| \langle \boldsymbol{v}, \boldsymbol{x} \rangle (v_s^k - v_r^k)\right) (1 + \mathcal{O}(1/N)).$$
$$(5.105)$$

Since $\boldsymbol{x} \in D^n(0, \delta_N)$ and $\exp(-ax)\sinh(ax)/ax = 1 - ax + \mathcal{O}(x^2)$, we obtain that to leading order the difference $g(\boldsymbol{x} - \hat{\boldsymbol{e}}_r^k + \hat{\boldsymbol{e}}_s^k) - g(\boldsymbol{x})$ is given by

$$g(\boldsymbol{x} - \hat{\boldsymbol{e}}^{k,r} + \hat{\boldsymbol{e}}^{k,s}) - g(\boldsymbol{x})$$

$$= (v_s^k - v_r^k) \sqrt{\tfrac{\beta|\hat{\gamma}_1|}{2\pi N}} \exp\left(-\tfrac{\beta N |\hat{\gamma}_1|}{2} \langle \boldsymbol{v}, \boldsymbol{x} \rangle^2\right) \left(1 - \tfrac{\beta|\hat{\gamma}_1|}{2} \langle \boldsymbol{v}, \boldsymbol{x} \rangle (v_s^k - v_r^k) + \mathcal{O}(\delta_N^2)\right).$$
$$(5.106)$$

Plugging (5.104), (5.105) and (5.106) into (5.103) and using that for all $x \in D^n(0, \delta_N)$

$$1 - \exp\left(\beta \langle \hat{e}^{k,s} - \hat{e}^{k,r}, (A + |\hat{\gamma}_1| v \cdot v^T) x \rangle\right)$$
$$= \beta \langle \hat{e}^{k,s} - \hat{e}^{k,r}, (A + |\hat{\gamma}_1| v \cdot v^T) x \rangle + \mathcal{O}(\delta_N^2),$$

we obtain, by collecting the leading order terms, that there exists $0 \leq c \leq \infty$ such that

$$\left|(\bar{L}^n g)(x)\right| \leq \sqrt{\frac{\beta|\hat{\gamma}_1|}{2\pi N}} \exp\left(-\tfrac{\beta N |\hat{\gamma}_1|}{2} \langle v, x \rangle^2\right) \left(c\delta_N^2 + \beta \langle v, R(A + |\hat{\gamma}_1| v \cdot v^T) x \rangle\right)$$
$$= \sqrt{\frac{\beta|\hat{\gamma}_1|}{2\pi N}} \exp\left(-\tfrac{\beta N |\hat{\gamma}_1|}{2} \langle v, x \rangle^2\right) c\delta_N^2. \tag{5.107}$$

Here, we used in the first step the definition of R, see (5.77). Let us point out that the second step relies on our choice of v and $\hat{\gamma}_1$ together with (5.82), (5.83) and the property (5.84). Namely,

$$\langle v, R(A + |\hat{\gamma}_1| v \cdot v^T) x \rangle = \sum_{i=1}^{n \cdot (q-1)} \hat{\gamma}_i \langle \check{v}^1, v^i \rangle \langle v^i, x \rangle - \hat{\gamma}_1 \langle \check{v}^1, v^i \rangle \langle v, x \rangle = 0. \tag{5.108}$$

This implies the statement of the lemma. □

After having shown that g is a good approximation of the equilibrium potential in a neighborhood of a critical point z, we can now proceed to compute an upper bound for the capacity.

Proposition 5.21. *For every $n \in \mathbb{N}$ and N sufficiently large we get*

$$\mathrm{cap}(A, B) \leq \mathcal{Q}^n(z) \frac{\beta|\hat{\gamma}_1|}{2\pi N} q^{-\frac{n}{2}} \prod_{i=1}^{n \cdot (q-1)} \sqrt{\frac{2\pi N}{\beta|\hat{\gamma}_i|}} \sqrt{\det(G^T G)} \left(1 + \mathcal{O}(\varepsilon(n) \wedge \delta_N)\right). \tag{5.109}$$

Proof. The strategy is to compute first the contribution to the mesoscopic Dirichlet form in the neighborhood $D^n(z, \delta_N)$ of the relevant saddle point z using the approximate harmonic function g. By exploiting Lemma 5.19, the computations will be done with the modification of the restricted Dirichlet form $\mathcal{E}_{D_N}^n$

$$\bar{\mathcal{E}}_{D_N}^n(g) := \sum_{x \in D^n} \sum_{k=1}^{n} \sum_{(r,s) \in \Delta_k} \bar{\mathcal{Q}}^n(x) \bar{r}^n(x, x - \hat{e}^{k,r} + \hat{e}^{k,s}) \left(g(x - \hat{e}^{k,r} + \hat{e}^{k,s}) - g(x)\right)^2. \tag{5.110}$$

Afterwards, we will show that the contribution to the Dirichlet form outside $D^n(z, \delta_N)$ is negligible. Again, we choose the origin of the coordinate system in such a way that $z = 0$, and we set $A \equiv A^n(z)$.

5.3. Upper bounds on capacities

First, note that by (5.106)

$$\left(g(\boldsymbol{x} - \hat{\boldsymbol{e}}^{k,r} + \hat{\boldsymbol{e}}^{k,s}) - g(\boldsymbol{x})\right)^2$$
$$= \frac{\beta |\hat{\gamma}_1|}{2\pi N} \exp\left(-\beta N |\hat{\gamma}_1| \langle \boldsymbol{v}, \boldsymbol{x} \rangle^2\right) \left(v_s^k - v_r^k\right)^2 \left(1 + \mathcal{O}(\delta_N)\right). \quad (5.111)$$

Inserting (5.111) together with the definitions of $\overline{\mathcal{Q}}^n$ and $\bar{r}^n$ into (5.110) gives

$$\overline{\mathcal{E}}_{D_N}^n(g) = \mathcal{Q}^n(0) \frac{\beta |\hat{\gamma}_1|}{2\pi N} \sum_{\boldsymbol{x} \in D^n} \exp\left(-\tfrac{\beta N}{2} \langle \boldsymbol{x}, (A + 2|\hat{\gamma}_1| \boldsymbol{v} \cdot \boldsymbol{v}^T) \boldsymbol{x} \rangle\right) \left(1 + \mathcal{O}(\delta_N)\right), \quad (5.112)$$

where we take additionally advantage of the fact that due to (5.82) and (5.83)

$$\sum_{k=1}^{n} \sum_{(r,s) \in \Delta_k} r_{r,s}^k \left(v_s^k - v_r^k\right)^2 = \langle \boldsymbol{v}, R\boldsymbol{v} \rangle = \langle \boldsymbol{v}^1, \check{\boldsymbol{v}}^1 \rangle = 1.$$

It remains to evaluate in (5.112) the sum over $\boldsymbol{x} \in D^n(0, \delta_N)$. By a standard approximation of the sum by an integral, we get that

$$\sum_{\boldsymbol{x} \in D^n} \exp\left(-\tfrac{\beta N}{2} \langle \boldsymbol{x}, (A + 2|\hat{\gamma}_1| \boldsymbol{v} \cdot \boldsymbol{v}^T) \boldsymbol{x} \rangle\right)$$
$$= \left(\frac{N^{q-1}}{\sqrt{q}}\right)^n \int_{V_1} \exp\left(-\tfrac{\beta N}{2} \langle \boldsymbol{x}, (A + 2|\hat{\gamma}_1| \boldsymbol{v} \cdot \boldsymbol{v}^T) \boldsymbol{x} \rangle\right) d\boldsymbol{x} \left(1 + \mathcal{O}(\sqrt{\ln N/N})\right). \quad (5.113)$$

Let us briefly explain the arising of the factor $(N^{q-1}/\sqrt{q})^n$. In order to replace the sum by an integral we have to take into account the volume of the unit cell associated to each lattice point that is given by the n-fold Cartesian product of the parallelepiped

$$\left\{ \tfrac{1}{N} \sum_{s=2}^{q} \lambda_s (e^s - e^r) \,\Big|\, \lambda_s \in [0, 1] \right\} \subset V_1 \subset \mathbb{R}^q,$$

of side length $1/N$. Its volume is given by $\sqrt{\det(I_{q-1} + 1 \cdot 1^T)} / N^{q-1} = \sqrt{q}/N^{q-1}$.

For the purpose of evaluating the integral, we consider the immersion

$$\mathbb{R}^{n \cdot (q-1)} \ni y \mapsto G \cdot (\hat{\boldsymbol{v}}^1, \ldots, \hat{\boldsymbol{v}}^{n \cdot (q-1)}) y \in \mathbb{R}^{n \cdot q},$$

where $\{\hat{\boldsymbol{v}}^1, \ldots, \hat{\boldsymbol{v}}^{n \cdot (q-1)}\}$ is an orthonormal basis of eigenvectors of $B^n(z)$. Then, in view of (5.82), we have

$$\int_{V_1} \exp\left(-\tfrac{\beta N}{2} \langle \boldsymbol{x}, (A + 2|\hat{\gamma}_1| \boldsymbol{v} \cdot \boldsymbol{v}^T) \boldsymbol{x} \rangle\right) d\boldsymbol{x}$$
$$= \prod_{k=1}^{n \cdot (q-1)} \int_{\mathbb{R}^{n(q-1)}} \exp\left(-\tfrac{\beta N}{2} |\hat{\gamma}_k| y^2\right) dy \sqrt{\det(G^T G)}$$
$$= \prod_{k=1}^{n \cdot (q-1)} \sqrt{\frac{2\pi}{\beta N |\hat{\gamma}_k|}} \sqrt{\det(G^T G)}, \quad (5.114)$$

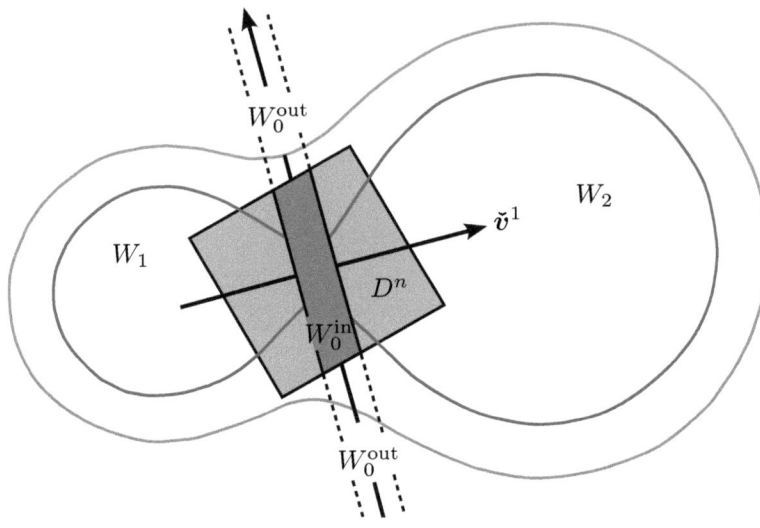

Figure 2. Domains for the construction of the test function $\tilde{g}$

where we used the non-orthogonal decomposition (5.84) in the second step. Therefore, we obtain the following estimate of the modified Dirichlet form $\bar{\mathcal{E}}^n_{D_N}$

$$\bar{\mathcal{E}}^n_{D_N}(g) = \mathcal{Q}^n(0) \frac{\beta|\hat{\gamma}_1|}{2\pi N} q^{-\frac{n}{2}} \prod_{k=1}^{n\cdot(q-1)} \sqrt{\frac{2\pi N}{\beta|\hat{\gamma}_k|}} \sqrt{\det\left(G^T G\right)} \left(1 + \mathcal{O}(\delta_N)\right). \quad (5.115)$$

It remains to show that the contribution to the Dirichlet form coming from points x outside of $D^n(z, \delta_N)$ do not contribute significantly to the capacity. Let us define the test function $\tilde{g}$ as

$$\tilde{g}(x) = \begin{cases} 0, & x \in W_1 \\ 1, & x \in W_2 \\ g(x), & x \in W_0 \end{cases}. \quad (5.116)$$

It is worth noting that the only non-zero contribution to the Dirichlet form $\mathcal{E}^n(\tilde{g})$ comes from $\overline{W}_0 := W_0 \cup \partial W_0$. Here, we denote by ∂W_0 the set of all points $x \notin W_0$ such that there exists $y \in W_0$ with $r^n(x, y) > 0$. Further, let us define the sets

$$W_0^{\text{in}} := W_0 \cap D^n \quad \text{and} \quad W_0^{\text{out}} := W_0 \cap (D^n)^c$$
$$\overline{W}_0^{\text{in}} := \overline{W}_0 \cap D^n \quad \text{and} \quad \overline{W}_0^{\text{out}} := W_0^{\text{out}} \cap \partial W_0^{\text{out}}.$$

5.3. Upper bounds on capacities

We denote by $\mathcal{E}^n_{\overline{W}_0^{\text{in}}}$ and $\mathcal{E}^n_{\overline{W}_0^{\text{out}}}$, respectively, the restriction of the Dirichlet form to the corresponding sets. Then,

$$\begin{aligned}\mathcal{E}^n(\tilde{g}) &= \overline{\mathcal{E}}^n_{\overline{W}_0^{\text{in}}}(\tilde{g})\left(1+\mathcal{O}(\max\{\varepsilon(n),\delta_N\})\right) + \mathcal{E}^n_{\overline{W}_0^{\text{out}}}(\tilde{g}) \\ &= \left(\overline{\mathcal{E}}^n_{\overline{W}_0^{\text{in}}}(g) - \left(\overline{\mathcal{E}}^n_{\overline{W}_0^{\text{in}}}(g) - \overline{\mathcal{E}}^n_{\overline{W}_0^{\text{in}}}(\tilde{g})\right)\right)\left(1+\mathcal{O}(\max\{\varepsilon(n),\delta_N\})\right) + \mathcal{E}^n_{\overline{W}_0^{\text{out}}}(\tilde{g})\end{aligned}$$
(5.117)

The remaining part of the proof comprises three steps.

STEP 1. Let us consider the term $\overline{\mathcal{E}}^n_{W_0^{\text{in}}}(g)$. For a suitable choosen $0 < c < 1$ such that $D^n(z, c\,\delta_N) \subseteq \overline{W}_0^{\text{in}}$ we have that

$$\overline{\mathcal{E}}^n_{D^n(z,c\,\delta_N)}(g) \leq \overline{\mathcal{E}}^n_{W_0^{\text{in}}}(g) \leq \overline{\mathcal{E}}^n_{D^n(z,\delta_N)}(g).$$

Hence, by inspecting the computation above, we obtain immediately that

$$\overline{\mathcal{E}}^n_{D^n(z,c\,\delta_N)}(g) = \overline{\mathcal{E}}^n_{D^n(z,\delta_N)}(g)\left(1+\mathcal{O}(\delta_N)\right).$$

Thus, it follows that

$$\overline{\mathcal{E}}^n_{W_0^{\text{in}}}(g) = \overline{\mathcal{E}}^n_{D^n(z,\delta_N)}(g)\left(1+\mathcal{O}(\delta_N)\right). \tag{5.118}$$

STEP 2. Consider now the term $\overline{\mathcal{E}}^n_{\overline{W}_0^{\text{in}}}(g) - \overline{\mathcal{E}}^n_{\overline{W}_0^{\text{in}}}(\tilde{g})$. Since, $\tilde{g} \equiv g$ on W_0, we get

$$\left|\overline{\mathcal{E}}^n_{\overline{W}_0^{\text{in}}}(g) - \overline{\mathcal{E}}^n_{\overline{W}_0^{\text{in}}}(\tilde{g})\right| \leq I_1 + I_2, \tag{5.119}$$

say, with

$$I_1 = \sum_{x \in W_1 \cap \partial W_0^{\text{in}}} \overline{\mathcal{Q}}^n(x) \sum_{\substack{1 \leq k \leq n \\ (r,s) \in \Delta_k}} r^k_{r,s} \left(\left(g(x) - g(x - \hat{e}^{k,r} + \hat{e}^{k,s})\right)^2 + g(x - \hat{e}^{k,r} + \hat{e}^{k,s})^2\right),$$

$$I_2 = \sum_{x \in W_2 \cap \partial W_0^{\text{in}}} \overline{\mathcal{Q}}^n(x) \sum_{\substack{1 \leq k \leq n \\ (r,s) \in \Delta_k}} r^k_{r,s} \left(\left(g(x) - g(x - \hat{e}^{k,r} + \hat{e}^{k,s})\right)^2 \right.$$

$$\left. + \left(1 - g(x - \hat{e}^{k,r} + \hat{e}^{k,s})\right)^2\right),$$

where we used that, by definition, the function $\tilde{g}$ has boundary condition zero and one, respectively, on W_1 and W_2. Due to the symmetry, we focus only on the first term. Notice that, for $x \in W_1 \cap \partial W_0^{\text{in}}$, it holds that $\langle x, v \rangle \leq -\delta_N$ and

$$g(x) \leq f(-\delta_N) \leq \frac{1}{\sqrt{2\pi\beta N |\hat{\gamma}_1|}\,\delta_N} \exp\left(-\frac{\beta N}{2}|\hat{\gamma}_1|\,\delta_N^2\right).$$

Inserting this estimate together with (5.106) into the expression for I_1 gives

$$I_1 \leq \frac{\beta|\hat{\gamma}_1|}{2\pi N} \exp\left(-\beta N |\hat{\gamma}_1| \delta_N^2\right) \sum_{x \in W_1 \cap \partial W_0^{\text{in}}} \overline{\mathcal{Q}}^n(x) \sum_{\substack{1 \leq k \leq n \\ (r,s) \in \Delta_k}} r_{r,s}^k \left((v_s^k - v_r^k)^2 + c\delta_N^{-2}\right)$$

$$= \mathcal{Q}^n(0) \frac{\beta|\hat{\gamma}_1|}{2\pi N} \exp\left(-\beta N |\hat{\gamma}_1| \delta_N^2\right) \sum_{x \in W_1 \cap \partial W_0^{\text{in}}} \exp\left(-\frac{\beta N}{2} \langle x, A x \rangle\right) c' \delta_N^{-2},$$

where we used in the second step that

$$\sum_{k=1}^n \sum_{(r,s) \in \Delta_k} r_{r,s}^k (v_s^k - v_r^k)^2 = \langle v, R v \rangle = \langle v^1, \check{v}^1 \rangle = 1.$$

Let us point out that the constant c' is independent of n and N. The evaluation of the remaining sum over $x \in W_1 \cap \partial W_0^{\text{in}}$ is similar to the computation we did above except for the fact that the integration runs over a $(n(q-1)-1)$-dimensional hyperplane W orthogonal to V_1 and shifted by the vector $\delta_n v$. Thus, setting $d = n(q-1)$, then for N large enough, we have

$$\sum_{x \in W_1 \cap \partial W_0^{\text{in}}} \exp\left(-\frac{\beta N}{2} \langle x, A x \rangle\right)$$

$$\leq \left(\frac{N^{q-1}}{\sqrt{q}}\right)^n \int_W \exp\left(-\frac{\beta N}{2} \langle x, A x \rangle\right) dx$$

$$= \left(\frac{N^{q-1}}{\sqrt{q}}\right)^n \exp\left(-\frac{\beta N}{2} \hat{\gamma}_1 \delta_N^2\right) \prod_{k=2}^{n(q-1)} \int_{\mathbb{R}^{d-1}} \exp\left(-\frac{\beta N}{2} |\hat{\gamma}_k| y^2\right) dy \sqrt{\det\left[G^T G\right]_d}$$

$$\leq \sqrt{N} \left(\frac{N^{q-1}}{\sqrt{q}}\right)^n \exp\left(\frac{\beta N}{2} |\hat{\gamma}_1| \delta_N^2\right) \prod_{k=2}^{n(q-1)} \sqrt{\frac{2\pi}{\beta N |\hat{\gamma}_k|}} \sqrt{\det(G^T G)}. \quad (5.120)$$

Notice that the second step relies on the non-orthogonal decomposition (5.84). Moreover, by taking advantage of the interlacing property of the eigenvalues of symmetric matrices and the fact that the eigenvalues of the matrix R_k are bounded from below by $c_{15}(n) \geq 1/N$, we bound the determinant of the principle minor $\det[G^T G]_{n(q-1)}$ by $N \det(G^T G)$. This implies that

$$I_1 \leq \mathcal{Q}^n(0) \frac{\beta|\hat{\gamma}_1|}{2\pi N} \prod_{k=2}^{n(q-1)} \sqrt{\frac{2\pi N}{\beta|\hat{\gamma}_k|}} \sqrt{\det(G^T G)} \, N \exp\left(-\frac{\beta N}{2} |\hat{\gamma}_1| \delta_N^2\right). \quad (5.121)$$

A similar bound can also be obtained for the term I_2. Finally, comparing these expression with (5.115) gives

$$\left|\bar{\mathcal{E}}_{W_0^{\text{in}}}^n(g) - \bar{\mathcal{E}}_{W_0^{\text{in}}}^n(\tilde{g})\right| \leq \bar{\mathcal{E}}_{D^n}(g) \, N \, e^{-\frac{1}{2}\beta|\hat{\gamma}_1|N^{2\delta}}. \quad (5.122)$$

STEP 3. It remains to consider the last Dirichlet form in (5.117). Since by assumption the saddle point is not degenerate and we are interested in an estimate orthogonal to the direction v, it holds for all $x \in \overline{W}_0^{\text{out}}$ that there exists $K < \infty$ such that $F^n(x) \geq F^n(z) + KN^{-1+2\delta}$ for some $K < \infty$ and N sufficiently large. Hence,

$$\mathcal{E}_{\overline{W}_0^{\text{out}}}^n(\tilde{g}) \leq \sum_{x \in \overline{W}_0^{\text{out}}} \mathcal{Q}^n(x) \leq c\,\mathcal{Q}^n(0)\,e^{-\beta KN^{2\delta}} \leq \overline{\mathcal{E}}_{D_N}^n(g)\,\mathcal{O}_N(\delta_N). \quad (5.123)$$

Combining (5.117) with the estimates given in (5.118), (5.122) and (5.123), yields that the Dirichlet form $\mathcal{E}^n(\tilde{g}) = \overline{\mathcal{E}}_{D_N}^n(g)\left(1 + \mathcal{O}(\max\{\varepsilon(n), \delta_N\})\right)$. This completes the proof of the upper bound. □

In view of Proposition 5.12, we obtain the following upper bound for capacities.

Corollary 5.22. *With the notation, introduced above, we obtain*

$$Z_N\,\text{cap}(A, B) \leq \frac{\beta|\bar{\gamma}_1|}{2\pi N}\,\frac{\exp(-\beta N\,F_N(z))}{\sqrt{|\det(I_q - 2\beta\,\text{Hess}\,U_N(2\beta z))|}}\,(1 + \mathcal{O}(\delta_N)). \quad (5.124)$$

Proof. First, recall that $|\hat{\gamma}_i|$ denote the eigenvalues of the matrix $B^n(z)$. Hence,

$$\prod_{k=1}^{n\cdot(q-1)} |\hat{\gamma}_k| = |\det(B^n(z))| = |\det(G^T A^n(z)G)| = \det(G^T G)\cdot|\det(A^n(z))|. \quad (5.125)$$

Substituting the expression (5.67) for the induced measure $\mathcal{Q}^n(z)$ at a critical point together with (5.68) and (5.125) into the upper bound (5.109) yields (5.124) except from the fact that instead of $\bar{\gamma}_1$ the eigenvalue $\hat{\gamma}_1$ of the matrix $B^n(z)$ appears. Notice that the error term $\mathcal{O}(\max\{\varepsilon(n), \delta_N\})$ becomes independent of n when n is chosen large enough. Hence, $\hat{\gamma}_1$ is the only n-dependent quantities on the right-hand side of (5.124) while the left-hand side is independent of n. By taking the limit $n \to \infty$ yields the desired bound. □

5.4. Lower bounds on capacities

In this section we will demonstrate how the Berman-Konsowa principle can be used to derive lower bounds on capacities. In view of Proposition 1.22, our task is to construct a suitable non-negative unit flow. The actual construction follows the strategy suggested in [6] and is done in two steps. As a first step, we construct a *mesoscopic unit flow* from the approximate harmonic function. In a second step, we construct for each mesoscopic path a *subordinate microscopic unit flow*. By a careful construction of both flows, we can establish a lower bound on the microscopic capacity in terms of the mesoscopic capacity that differs only by a factor of size $1 + \mathcal{O}(\varepsilon(n))$.

5.4.1. Mesoscopic and microscopic lower bounds: The strategy.

Let us consider two minima, m, m', of the mesoscopic free energy F^n. Further, let us denote by z the lowest saddle point of F^n between m and m', i.e. $F^n(\hat{z}) = \min_\gamma \max_{x \in \gamma} F^n(x)$ where the minimum is taken over all mesoscopic paths γ on Γ^n between $\hat{m}$ and $\hat{m}'$. Here, $\hat{x} \in \Gamma^n$ denotes the closest lattice point approximation of $m \in \times_{k=1}^n \mathcal{M}_{\pi_k}$. For convenience, we pretend that $m, z, m' \in \Gamma^n$, since the proofs will not be sensitive concerning this correction. As in the previous section, let $A, B \subset \Gamma^n$ be two subsets which are strictly contained in two different connected components of the level sets $\{x \in \Gamma^n \mid F^n(x) < F^n(z)\}$ such that $m \in A$ and $m' \in B$.

In view of (1.85), we will construct a non-negative unit flow $\mathfrak{f}_{A,B}$ of the form

$$\mathfrak{f}_{A,B}(x,y) = \frac{Q^n(x)\,r^n(x,y)}{\mathcal{E}^n(\tilde{g})} \phi_{A,B}(x,y), \qquad (5.126)$$

such that with respect to the law, $\mathbb{P}^{\mathfrak{f}_{A,B}}$, of the associated Markov chain it holds

$$\mathbb{P}^{\mathfrak{f}_{A,B}}\left[\sum_{(x,y)\in\gamma} \phi_{A,B}(x,y) = 1 + o_N(1)\right] = 1 - o_N(1). \qquad (5.127)$$

As an immediate consequence of Proposition 1.22, Equation (5.127) implies that the mesoscopic capacity is bounded from below by

$$\mathrm{CAP}^n(A,B) \geq \mathbb{E}^{\mathfrak{f}_{A,B}}\left[\left(\sum_{(x,y)\in\gamma} \frac{\mathfrak{f}_{A,B}(x,y)}{Q^n(x)\,r^n(x,y)}\right)^{-1}\right] \geq \mathcal{E}^n(\tilde{g})\left(1 - o_N(1)\right). \qquad (5.128)$$

As demonstrated in [6], for a suitable chosen neighborhood $\mathfrak{D}^n$ of the saddle point z, the flow $\mathfrak{f}_{A,B}$ consists of a concatenation of a flow $\mathfrak{f}_A$ from A to the boundary $\partial_A \mathfrak{D}^n$, a flow $\mathfrak{f}$ through $\mathfrak{D}^n$ and a flow $\mathfrak{f}_B$ from $\partial_B \mathfrak{D}^n$ to B. It turns out that, with a probability close to one, the major contribution to the sum in (5.127) comes from the flow $\mathfrak{f}$, i.e.

$$\mathbb{P}^{\mathfrak{f}}\left[\sum_{(x,y)\in\gamma} \phi(x,y) = 1 + o_N(1)\right] = 1 - o_N(1).$$

The actual construction of $\mathfrak{f}$, that is presented in the next subsection, is the most difficult part and relies on the following observation. Although the test function g, which we have constructed in (5.97), does not satisfy Kirchhoff's law, Lemma 5.20 suggests that inside $\mathfrak{D}^n$ a potential candidate for ϕ seems to be a (small) perturbation of the discrete gradient ∇g. In contrast to that, we will show that with a probability close to one the contribution to (5.127) coming from $\mathfrak{f}_A$ and $\mathfrak{f}_B$ is indeed negligible.

After having established a mesoscopic lower bound via a suitable mesoscopic flow, our next task is to construct a subordinate *microscopic flow*, $f_{A,B}$, from $A = \mathcal{S}^n[A]$ to $B = \mathcal{S}^n[B]$. In order to do so, let $\mathbf{X} = \{\mathbf{X}(t)\}_t$ be the Markov chain on Γ^n with law $\mathbb{P}^{\mathfrak{f}_{A,B}}$ that starts in A and is stopped on the arrival of B. Given a realization x of this mesoscopic Markov chain, we choose the label in such a way that

5.4. Lower bounds on capacities

$\underline{x} = (x(a_{\underline{x}}), \ldots, x(0), \ldots, x(b_{\underline{x}}))$ with $x(a_{\underline{x}}) \in A$, $x(b_{\underline{x}}) \in B$ and $\sum_{k=1}^{n} x(0)^k = \sum_{k=1}^{n} z^k$. Notice that $a_{\underline{x}} < 0$ and $b_{\underline{x}} > 0$. To each path, $\underline{x}$, of positive probability associate a subordinate microscopic unit flow $f^{\underline{x}}$ such that

$$f^{\underline{x}}(\sigma, \eta) > 0 \quad \Longleftrightarrow \quad (\varrho^n(\sigma), \varrho^n(\eta)) \in \underline{x}. \tag{5.129}$$

Hence, the total microscopic flow $f_{A,B}$ can be decomposed as

$$f_{A,B}(\sigma, \eta) = \sum_{\underline{x}} \mathbb{P}^{f_{A,B}}[\underline{X} = \underline{x}] \, f^{\underline{x}}(\sigma, \eta). \tag{5.130}$$

Due to the fact that $f^{\underline{x}}$ is a unit flow, it holds that

$$\sum_{\sigma \in S^n[x]} \sum_{\eta \in S^n[y]} f^{\underline{x}}(\sigma, \eta) = 1, \quad \forall (x, y) \in \underline{x}. \tag{5.131}$$

As an immediate consequence of (5.130) and (5.131), the total microscopic flow has the property

$$\sum_{\sigma \in S^n[x]} \sum_{\eta \in S^n[y]} f_{A,B}(\sigma, \eta) = \sum_{\underline{x}} \mathbb{P}^{f_{A,B}}[\underline{X} = \underline{x}] \, \mathbb{1}_{(x,y) \in \underline{x}} = \mathfrak{f}_{A,B}(x, y). \tag{5.132}$$

Further, for each subordinate microscopic flow let us associate a microscopic Markov chain $\Sigma^{\underline{x}} = \{\Sigma^{\underline{x}}(t)\}_t$ on $\mathcal{S}_N$ with law $\mathbb{P}^{f^{\underline{x}}}$. Notice that (5.130) combined with (5.132) give rise to the following decomposition of unity

$$\mathbb{1}_{f_{A,B}(\varrho^n(\sigma), \varrho^n(\eta)) > 0} = \sum_{\underline{x}} \sum_{\underline{x} \ni (\sigma, \eta)} \frac{\mathbb{P}^{f_{A,B}}[\underline{X} = \underline{x}] \, \mathbb{P}^{f^{\underline{x}}}[\Sigma^{\underline{x}} = \sigma]}{\mathfrak{f}_{A,B}(\varrho^n(\sigma), \varrho^n(\eta)) \, f^{\underline{x}}(\sigma, \eta)}. \tag{5.133}$$

Hence, by Proposition 1.22, the capacity $\mathrm{cap}(A, B)$ is bounded from below by

$$\mathrm{cap}(A, B)$$

$$\geq \sum_{\underline{x}} \mathbb{P}^{f_{A,B}}[\underline{X} = \underline{x}] \, \mathbb{E}^{f^{\underline{x}}} \left[\left(\sum_{t=a_{\underline{x}}}^{b_{\underline{x}}-1} \frac{\mathfrak{f}_{A,B}(x(t), x(t+1)) \, f^{\underline{x}}(\sigma(t), \sigma(t+1))}{\mu_N(\sigma(t)) \, p_N(\sigma(t), \sigma(t+1))} \right)^{-1} \right]$$

$$\geq \mathcal{E}^n(\tilde{g}) \sum_{\underline{x}} \mathbb{P}^{f_{A,B}}[\underline{X} = \underline{x}] \left(\sum_{t=a_{\underline{x}}}^{b_{\underline{x}}-1} \mathbb{E}^{f^{\underline{x}}}[\Phi_t^{\underline{x}}] \, \phi_{A,B}(x(t), x(t+1)) \right)^{-1} \tag{5.134}$$

with

$$\Phi_t^{\underline{x}}(\varrho) = \frac{\mathcal{Q}^n(x(t)) \, r^n(x(t), x(t+1))}{\mu_N(\sigma(t)) \, p_N(\sigma(t), \sigma(t+1))} \, f^{\underline{x}}(\sigma(t), \sigma(t+1)) \tag{5.135}$$

where we applied in the last line Jensen's inequality and replaced $\mathfrak{f}_{A,B}$ by the expression (5.126). Hence, it remains to construct for each macroscopic path $\underline{x}$ a subordinate unit flow $f^{\underline{x}}$.

Remark 5.23. Let us consider for a moment the artificial case where all $\tilde{h}_i \equiv 0$. In view of (5.79), it holds that $\sum_{\eta \in \mathcal{S}^n[\boldsymbol{x}(t+1)]} p_N(\sigma, \eta) = r^n(\boldsymbol{x}(t), \boldsymbol{x}(t+1))$ for all $\sigma \in \mathcal{S}^n[\boldsymbol{x}(t)]$. Therefore, we can define

$$f^{\boldsymbol{x}}(\sigma(t), \sigma(t+1)) := \frac{\mu_N(\sigma(t)) \, p_N(\sigma(t), \sigma(t+1))}{\mathcal{Q}^n(\boldsymbol{x}(t)) \, r^n(\boldsymbol{x}(t), \boldsymbol{x}(t+1))}$$

which is indeed a unit flow, because Kirchhoff's law is satisfied, i.e.

$$\sum_{\eta \in \mathcal{S}^n[\boldsymbol{x}(t-1)]} f^{\boldsymbol{x}}(\eta, \sigma(t)) = \frac{\mu_N(\sigma(t))}{\mathcal{Q}^n(\boldsymbol{x}(t))} = \frac{1}{|\mathcal{S}^n[\boldsymbol{x}(t)]|} = \sum_{\eta \in \mathcal{S}^n[\boldsymbol{x}(t+1)]} f^{\boldsymbol{x}}(\sigma(t), \eta).$$

Note, that this choice of a subordinate flow implies that $\Phi_t^{\boldsymbol{x}}(\sigma) \equiv 1$ for all $t = a_{\boldsymbol{x}}, \ldots, b_{\boldsymbol{x}}$. Thus it is immediate that $\mathrm{cap}(A, B) \geq \mathcal{E}^n(\tilde{g})$.

In view of the assumption on the distribution of the magnetic field and the coarse graining procedure, there is some hope to construct a subordinate flow $f^{\boldsymbol{x}}$ such that $\Phi_t^{\boldsymbol{x}}(\sigma)$ is very close to one. Due to the fact that we are interested in proving a lower bound, it is enough to consider a subset of all realizations of the mesoscopic chain. This subset is chosen in such a way that its $\mathbb{P}^{\mathfrak{f}_A,B}$-probability tends to one as $N \uparrow \infty$, and that we are able to construct a subordinate flow such that uniformly for all those realizations $\boldsymbol{x}$ it holds

$$\sum_{t=a_{\boldsymbol{x}}}^{b_{\boldsymbol{x}}-1} \mathbb{E}^{f^{\boldsymbol{x}}} \left[\Phi_t^{\boldsymbol{x}} \right] \phi_{A,B}(\boldsymbol{x}(t), \boldsymbol{x}(t+1)) \leq 1 + \mathcal{O}(\epsilon(n)).$$

Clearly, if we are able to find such a subset, we immediately obtain

$$\mathrm{cap}(A, B) \geq \mathcal{E}^n(\tilde{g}) \left(1 - \mathcal{O}(\epsilon(n))\right).$$

5.4.2. Construction of the mesoscopic flow. Let us point out that the construction of $\mathfrak{f}$ and $\mathfrak{f}_A, \mathfrak{f}_B$ presented in this subsection is an adaptation of the corresponding construction given in [6]. Our first aim is to construct a flow $\mathfrak{f}$ on Γ^n within the set

$$\mathfrak{D}^n \equiv \mathfrak{D}^n(\boldsymbol{z}, \nu\delta_N) := D^n(\boldsymbol{z}, \delta_N) \cap \{\boldsymbol{x} \in \Gamma^n \mid |\langle \boldsymbol{x} - \boldsymbol{z}, \check{v} \rangle| < \nu\delta_N\}, \qquad (5.136)$$

for some small number $\nu > 0$. Here, $\check{v} \equiv \check{v}^1$ is the vector previously defined in (5.82). Note that the boundary of $\mathfrak{D}^n$, denoted by $\partial \mathfrak{D}^n$, consists of three disjoint parts. Namely,

$$\partial \mathfrak{D}^n = \partial_A \mathfrak{D}^n \cup \partial_B \mathfrak{D}^n \cup \partial_r \mathfrak{D}^n, \qquad (5.137)$$

where $\partial_A \mathfrak{D}^n := \{\boldsymbol{x} \in \partial \mathfrak{D}^n \mid \langle \boldsymbol{x} - \boldsymbol{z}, \check{v} \rangle \leq -\nu\delta_N\}$ and $\partial_B \mathfrak{D}^n := \{\boldsymbol{x} \in \partial \mathfrak{D}^n \mid \langle \boldsymbol{x} - \boldsymbol{z}, \check{v} \rangle \geq \nu\delta_N\}$, while $\partial \mathfrak{D}_r^n$ denotes all points $\boldsymbol{x}$ at the boundary of $D^n(\boldsymbol{z}, \delta_N)$ such that $|\langle \boldsymbol{x} - \boldsymbol{z}, \check{v} \rangle| < \nu\delta_N$, see Figure 5.4.2. We choose ν small enough to guarantee that there exists $K > 0$ such that

$$F^n(\boldsymbol{x}) \geq F^n(\boldsymbol{z}) + K\delta_N^2, \qquad \forall \boldsymbol{x} \in \partial_r \mathfrak{D}^n. \qquad (5.138)$$

5.4. Lower bounds on capacities

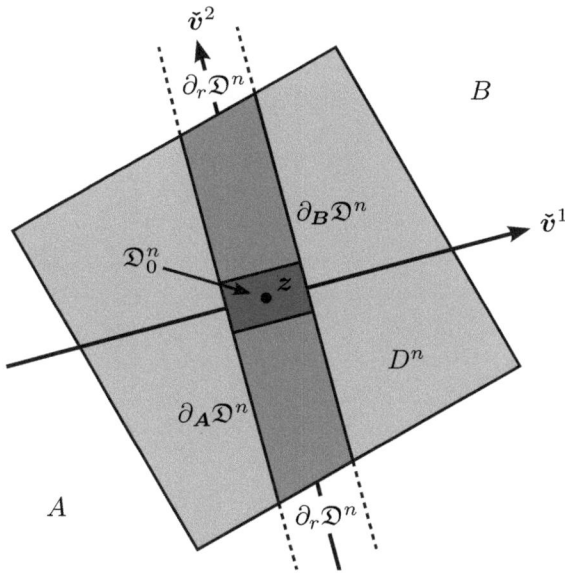

Figure 3. Sketch of the set $D^n(z, \delta_N)$ and the subset $\mathfrak{D}^n$ that is used in the construction of the flow $\mathfrak{f}$. Notice that the set $\mathfrak{D}_0^n$ is a narrow tube along the direction $\check{v}^1$ lying inside $\mathfrak{D}^n$.

Having in mind the definition of the harmonic flow (1.85) as well as Lemma 5.20, we will construct a non-negative unit flow $\mathfrak{f}$ of the form

$$\mathfrak{f}(x, x - \hat{e}^{k,r} + \hat{e}^{k,s}) := \frac{c(N)}{\mathcal{E}^n(\tilde{g})} \mathcal{F}[\phi](x, x - \hat{e}^{k,r} + \hat{e}^{k,s}), \qquad (5.139)$$

for all $k = 1, \ldots, n$ and $(r, s) \in \Delta_k$ while $\mathfrak{f} \equiv 0$ otherwise. Here we introduced the notation

$$\mathcal{F}[\phi](x, x - \hat{e}^{k,r} + \hat{e}^{k,s}) = \overline{Q}^n(x)\, \bar{r}^n(x, x - \hat{e}^{k,r} + \hat{e}^{k,s})\, \phi(x, x - \hat{e}^{k,r} + \hat{e}^{k,s}). \qquad (5.140)$$

Notice that $c(N) = 1 + o_N(1)$ is a suitable normalization constants.

Hence, our task is to specify a function ϕ of the form $\phi = \phi_T^0 + u$ where ϕ_T^0 is the truncation of a suitable function ϕ^0 and u should be a small correction. A natural candidate for ϕ^0 would be the discrete gradient of the test function g, see (5.97), because g is almost harmonic as shown in Lemma 5.20. However, due to the fact that we have to add a correction u anyway in order to satisfy Kirchhoff's law, it is more convenient to start with

$$\phi^0(\boldsymbol{x}, \boldsymbol{x} - \hat{e}^{k,r} + \hat{e}^{k,s}) := (v_r^k - v_s^k)\sqrt{\frac{\beta|\hat{\gamma}_1|}{2\pi N}}\exp\left(-\frac{\beta N|\hat{\gamma}_1|}{2}\langle \boldsymbol{x} - \boldsymbol{z}, \boldsymbol{v}\rangle^2\right) \quad (5.141)$$

for all $k = 1, \ldots, n$ and $(r, s) \in \Delta_k$. In order to construct the correction u we proceed as follows. First, fix $0 < \nu_0 \ll \nu$ small enough and define a narrow tube $\mathfrak{D}_0^n$ along the direction $\check{\boldsymbol{v}}$, i.e.

$$\mathfrak{D}_0^n \equiv \mathfrak{D}_0^n(\boldsymbol{z}, \nu_0\delta_N) := \mathfrak{D}^n \cap \left\{\boldsymbol{x} \in \Gamma^n \mid \left\|\boldsymbol{x} - \boldsymbol{z} - \langle \boldsymbol{x} - \boldsymbol{z}, \check{\boldsymbol{v}}\rangle\frac{\check{\boldsymbol{v}}}{\|\check{\boldsymbol{v}}\|_2^2}\right\|_2 < \nu_0\delta_N\right\}, \quad (5.142)$$

and let $\mathcal{C}$ be the cone spanned by the vectors $\{e^{k,s} - e^{k,r} \mid k = 1, \ldots, n, (r, s) \in \Delta_k\}$. Mind that, under the above Assumption 5.18, the vector v lies in the interior of $\mathcal{C}$. Further, we define

$$\mathfrak{D}_1^n := \{\boldsymbol{x} - \mathcal{C} \mid \boldsymbol{x} \in \partial_B \mathfrak{D}_0^n\} \cap \mathfrak{D}^n, \quad \mathfrak{D}_2^n := \{\boldsymbol{x} + \mathcal{C} \mid \boldsymbol{x} \in \partial_A \mathfrak{D}_1^n\} \cap \mathfrak{D}^n, \quad (5.143)$$

where $\partial_A \mathfrak{D}_1^n$ denotes all points $\boldsymbol{x}$ at the boundary $\partial \mathfrak{D}^n$ such that $\boldsymbol{x} - \hat{e}^{k,r} + \hat{e}^{k,s} \in \mathfrak{D}_1^n$ for all $k = 1, \ldots, n$ and $(r, s) \in \Delta_k$. We assume that the constants ν and ν_0 in the definitions of $\mathfrak{D}^n$ and $\mathfrak{D}_0^n$ are such that $\mathfrak{D}_2^n \cap \partial_r \mathfrak{D}^n = \emptyset$. Moreover, we denote by ϕ_T^0 the restriction of ϕ^0 to $\mathfrak{D}_1^n \cup \partial_A \mathfrak{D}_1^n$

$$\phi_T^0(\boldsymbol{x}, \boldsymbol{x} - \hat{e}^{k,r} + \hat{e}^{k,s}) = \phi^0(\boldsymbol{x}, \boldsymbol{x} - \hat{e}^{k,r} + \hat{e}^{k,s})\mathbb{1}_{\boldsymbol{x} \in \mathfrak{D}_1^n \cup \partial_A \mathfrak{D}_1^n}. \quad (5.144)$$

Remark 5.24. Let us now explain briefly the idea behind the construction of the sets $\mathfrak{D}_1^n$ and $\mathfrak{D}_2^n$. Consider a point $\boldsymbol{x} \in \mathfrak{D}_1^n$. In order to satisfy Kirchhoff's law at $\boldsymbol{x}$, the incoming flow should equal the outgoing flow, i.e. $\sum_{\boldsymbol{y} \in \Gamma^n} \mathfrak{f}(\boldsymbol{y}, \boldsymbol{x}) = \sum_{\boldsymbol{y} \in \Gamma^n} \mathfrak{f}(\boldsymbol{x}, \boldsymbol{y})$. Recall that for a non-negative flow $\mathfrak{f}(\boldsymbol{x}, \boldsymbol{y}) > 0$ implies $\mathfrak{f}(\boldsymbol{y}, \boldsymbol{x}) = 0$. By construction of $\mathfrak{D}_1^n$ and ϕ_T^0, it holds that $\{\boldsymbol{y} \in \Gamma^n \mid \mathfrak{f}(\boldsymbol{y}, \boldsymbol{x}) > 0\} \subset \mathfrak{D}_1^n$. Hence, the incoming flow arriving at $\boldsymbol{x}$ can only come from points in $\mathfrak{D}_1^n$ whereas points in $\mathfrak{D}_2^n$ cannot contribute to the incoming flow to $\boldsymbol{x}$. Let us stress the fact that this property holds true for all $\boldsymbol{x} \in \mathfrak{D}_1^n$. Hence, the total mass accumulated at $\partial_B \mathfrak{D}_0^n$ can only come from $\partial_A \mathfrak{D}_1^n$. For this reason, we will distribute the initial mass one only among the points in $\partial_A \mathfrak{D}_1^n$.

On the other hand, concerning the outgoing flow, a mass transport from $\mathfrak{D}_1^n$ to $\mathfrak{D}_2^n \setminus \mathfrak{D}_1^n$ is possible for all points $\boldsymbol{x} \in \mathfrak{D}_1^n$ and $\boldsymbol{y} \in \mathfrak{D}_2^n$ such that $r^n(\boldsymbol{x}, \boldsymbol{y}) > 0$. Hence, while transporting the initial total mass one from $\partial_A \mathfrak{D}_1^n$ to $\partial_B \mathfrak{D}_0^n$ we may loose some mass. The reason behind the truncation of ϕ^0 to $\mathfrak{D}_1^n$ is to minimize these losses. Further, since the total mass can only be transported to $\partial_B \mathfrak{D}_2^n$ it is enough to construct the correction u on $\mathfrak{D}_2^n$.

5.4. Lower bounds on capacities

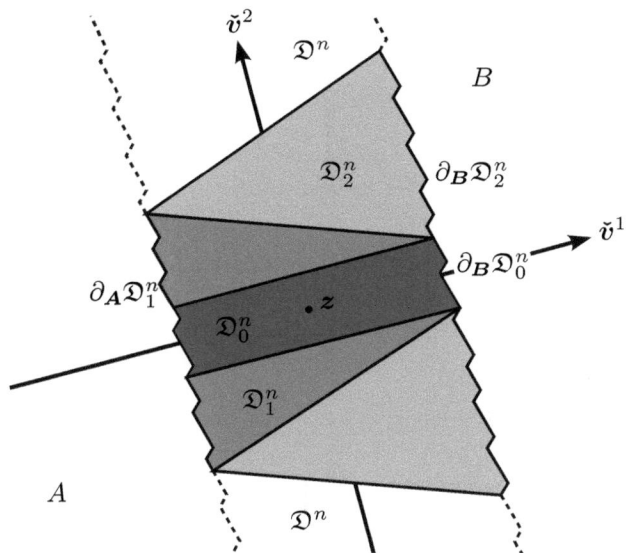

Figure 4. Sketch of the narrow tube $\mathfrak{D}_0^n$ along the direction $\check{v}^1$ together with the sets $\mathfrak{D}_1^n$ and $\mathfrak{D}_2^n$.

Now, we turn to the construction of the function u. In order to satisfy Kirchhoff's law inside $\mathfrak{D}^n$, the function u should satisfy the following recursion

$$\sum_{\substack{1\leq k\leq n \\ (r,s)\in\Delta_k}} \mathcal{F}[u](x, x - \hat{e}^{k,r} + \hat{e}^{k,s}) = \sum_{\substack{1\leq k\leq n \\ (r,s)\in\Delta_k}} \mathcal{F}[u](x + \hat{e}^{k,r} - \hat{e}^{k,s}, x) - \mathrm{d}\mathcal{F}[\phi_T^0](x) \tag{5.145}$$

where

$$\mathrm{d}\mathcal{F}[\phi_T^0](x) := \sum_{\substack{1\leq k\leq n \\ (r,s)\in\Delta_k}} \left(\mathcal{F}[\phi_T^0](x, x - \hat{e}^{k,r} + \hat{e}^{k,s}) - \mathcal{F}[\phi_T^0](x + \hat{e}^{k,r} - \hat{e}^{k,s}, x) \right).$$

Since $\phi_T^0 \equiv 0$ on $\mathfrak{D}^n \setminus \mathfrak{D}_1^n$ and due to the construction of $\mathfrak{D}_2^n$, we may set $u \equiv 0$ on $\mathfrak{D}^n \setminus \mathfrak{D}_2^n$. In the sequel, we solve the recursion (5.145) explicitly inside $\mathfrak{D}_1^n$. For this purpose, we slice $\mathfrak{D}_1^n$ into layers $\mathfrak{L}_k$ in the following way. Namely, set $\mathfrak{L}_0 := \partial_A \mathfrak{D}_1^n$ and for $j \in \mathbb{N}_0$ define

$$\mathfrak{L}_{j+1} := \left\{ x \in \mathfrak{D}^n \,\middle|\, x + \hat{e}^{k,r} - \hat{e}^{k,s} \in \bigcup_{i=1}^{j} \mathfrak{L}_i, \ \forall k = 1, \ldots, n, \ (r,s) \in \Delta_k \right\}. \tag{5.146}$$

Since v lies inside the cone $\mathcal{C}$, there exists $c_{19}(n) > 0$ and $T \leq c_{19}(n)N\delta_N$ such that

$$\mathfrak{D}_1^n = \bigcup_{j=0}^{T} \mathfrak{L}_j.$$

By setting $u \equiv 0$ on $\partial_A \mathfrak{D}^n$, we define recursively, for every $x \in \mathfrak{L}_{j+1}$

$$\mathcal{F}[u](x, x - \hat{e}^{k,r} + \hat{e}^{k,s}) := q_{r,s}^k \Bigg(\sum_{\substack{1 \leq l \leq n \\ (r',s') \in \Delta_k}} \mathcal{F}[u](x + \hat{e}^{l,r'} - \hat{e}^{l,s'}, x) - \mathrm{d}\mathcal{F}[\phi_T^0](x) \Bigg), \tag{5.147}$$

where the probability distribution $\{q_{r,s}^k \mid k = 1, \ldots, n, \ (r,s) \in \Delta_k\}$ is defined by

$$q_{r,s}^k = \frac{r_{r,s}^k \left(v_r^k - v_s^k \right)}{\sum_{l=1}^n \sum_{(t,u) \in \Delta_k} r_{t,u}^l \left(v_t^l - v_u^l \right)}. \tag{5.148}$$

Obviously, this produces a solution to (5.145) on $\mathfrak{D}_1^n$ whereas on $\mathfrak{D}_2^n \setminus \mathfrak{D}_1^n$ u is implicitly given by the solution of (5.145) subject to additional boundary conditions on $\partial_r \mathfrak{D}_2^n$ and $\partial_r \mathfrak{D}_1^n$. Let us emphasis the fact that the positivity of $q_{r,s}^k$ relies crucially on the Assumption 5.18.

Proposition 5.25. *The flow $\mathfrak{f}$, as constructed above, satisfies the following properties:*

(i) *For all $x \in \mathfrak{D}^n$ it holds that*

$$\mathrm{d}\mathcal{F}[\phi](x) = 0 \quad \text{and} \quad \sum_{x \in \partial_A \mathfrak{D}^n} \sum_{(r,s) \in \Delta_k} \mathfrak{f}(x, x - \hat{e}^{k,r} + \hat{e}^{k,s}) = 1. \tag{5.149}$$

(ii) *Uniformly for all $x \in \mathfrak{D}_0^n$ it holds that*

$$\phi(x, x - \hat{e}^{k,r} + \hat{e}^{k,s}) = \big(g(x) - g(x, x - \hat{e}^{k,r} + \hat{e}^{k,s})\big)(1 + o_N(1)), \tag{5.150}$$

for all $k = 1, \ldots, n$ and $(r,s) \in \Delta_k$.

(iii) *There exists $\kappa > 0$ such that*

$$\max_{x \in \mathfrak{D}^n \setminus \mathfrak{D}_0^n} \max_{k=1,\ldots,n} \max_{(r,s) \in \Delta_k} \mathfrak{f}(x, x - \hat{e}^{k,r} + \hat{e}^{k,s}) \leq N^{-\kappa}. \tag{5.151}$$

Proof. (i) As an immediate consequence of the construction of ϕ, the first statement in (5.149) is satisfied for all $x \in \mathfrak{D}^n$, i.e. $\mathfrak{f}$ is a non-negative, cycle-free flow. It remains to prove that $\mathfrak{f}$ is a unit flow.

5.4. Lower bounds on capacities

By inspecting the proof of Proposition 5.21 we see that $\mathcal{E}^n(\tilde{g}) = \bar{\mathcal{E}}^n_{\mathfrak{D}^n \cup \partial_A \mathfrak{D}^n}(g)(1 + \mathcal{O}_N(1))$. On the other hand, Green's first identify implies that

$$\bar{\mathcal{E}}^n_{\mathfrak{D}^n \cup \partial_A \mathfrak{D}^n}(g) + \sum_{x \in \mathfrak{D}^n} \bar{Q}^n(x) \, g(x) \, (\bar{L}^n g)(x)$$
$$= -\sum_{\substack{x \in \partial_B \mathfrak{D}^n \\ (r,s) \in \Delta_k(x)}} \sum_{1 \le k \le n} \bar{Q}^n(x) \, g(x) \, \bar{r}^n(x, x + \hat{e}^{k,r} - \hat{e}^{k,s}) \left(g(x + \hat{e}^{k,r} - \hat{e}^{k,s}) - g(x) \right)$$
$$- \sum_{\substack{x \in \partial_A \mathfrak{D}^n \\ (r,s) \in \Delta_k(x)}} \sum_{1 \le k \le n} \bar{Q}^n(x) \, g(x) \, \bar{r}^n(x, x - \hat{e}^{k,r} + \hat{e}^{k,s}) \left(g(x - \hat{e}^{k,r} + \hat{e}^{k,s}) - g(x) \right),$$
(5.152)

where, for $x \in \partial_A \mathfrak{D}^n$ and $k \in \{1, \ldots, n\}$, $\Delta_k(x)$ denotes the subset of all $(r, s) \in \Delta_k$ such that $x - \hat{e}^{k,r} + \hat{e}^{k,s} \in \mathfrak{D}^n$ and analog for $x \in \partial_B \mathfrak{D}^n$. Recall that the function g converge exponentially fast in N to 0 for all $x \in \partial_B \mathfrak{D}^n$ and to 1 for all $x \in \partial_A \mathfrak{D}^n$. Hence, the first term on the right-hand side of (5.152) is negligible and, in the second term, we can replace $g(x)$ by one. Moreover, a comparison of (5.102) with (5.111) reveals that

$$\left| \sum_{x \in \mathfrak{D}^n} \bar{Q}^n(x) \, g(x) \, (\bar{L}^n g)(x) \right| = \bar{\mathcal{E}}^n_{\mathfrak{D}^n \cup \partial_A \mathfrak{D}^n}(g) \left(1 + \mathcal{O}(N^{-1/2+2\delta}) \right). \quad (5.153)$$

Hence,

$$\bar{\mathcal{E}}^n_{\mathfrak{D}^n \cup \partial_A \mathfrak{D}^n}(g) \left(1 + \mathcal{O}_N(1) \right) = -\sum_{x \in \partial_A \mathfrak{D}^n} \sum_{k=1}^n \sum_{(r,s) \in \Delta_k(x)} \mathcal{F}[\nabla g](x, x - \hat{e}^{k,r} + \hat{e}^{k,s}). \quad (5.154)$$

Here, we introduced the discrete gradient, ∇g, that is defined by $(\nabla g)(x, y) := g(y) - g(x)$. By employing the construction of ϕ we infer that

$$\sum_{\substack{x \in \partial_A \mathfrak{D}^n \\ (r,s) \in \Delta_k(x)}} \sum_{1 \le k \le n} \mathcal{F}[\phi](x, x - \hat{e}^{k,r} + \hat{e}^{k,s}) = \sum_{\substack{x \in \partial_A \mathfrak{D}^n_1 \\ (r,s) \in \Delta_k(x)}} \sum_{1 \le k \le n} \mathcal{F}[\phi^0](x, x - \hat{e}^{k,r} + \hat{e}^{k,s}).$$

Notice that by (5.106)

$$\phi^0(x, x - \hat{e}^{k,r} + \hat{e}^{k,s}) = \left(g(x) - g(x - \hat{e}^{k,r} + \hat{e}^{k,s}) \right) (1 + \mathcal{O}(\delta_N)) \quad (5.155)$$

uniformly for all $x \in \mathfrak{D}^n$. Therefore, we are left with showing that the contribution coming from $\partial_A \mathfrak{D}^n \setminus \partial_A \mathfrak{D}^n_1$ on the right-hand side of (5.154) is small. By using that

$$-\mathcal{F}[\nabla g](x, x - \hat{e}^{k,r} + \hat{e}^{k,s}) = \mathcal{F}[\phi^0](x, x - \hat{e}^{k,r} + \hat{e}^{k,s}) (1 + \mathcal{O}(\delta_N))$$

together with the non-orthogonal decomposition (5.84) we have that

$$\frac{\mathcal{F}[\phi^0]\left(\boldsymbol{x}, \boldsymbol{x} - \hat{\boldsymbol{e}}^{k,r} + \hat{\boldsymbol{e}}^{k,s}\right)}{\mathcal{Q}^n(\boldsymbol{z})\, r_{r,s}^k\left(v_r^k - v_s^k\right)}$$

$$= \sqrt{\frac{\beta|\hat{\gamma}_1|}{2\pi N}} \exp\left(-\tfrac{\beta N}{2} \left\langle \boldsymbol{x} - \boldsymbol{z}, \left(A^n(\boldsymbol{z}) + |\hat{\gamma}_1|\boldsymbol{v} \cdot \boldsymbol{v}^{\mathrm{T}}\right)(\boldsymbol{x} - \boldsymbol{z})\right\rangle\right)$$

$$= \sqrt{\frac{\beta|\hat{\gamma}_1|}{2\pi N}} \exp\left(-\tfrac{\beta N}{2} \sum_{i=2}^{n\cdot(q-1)} \hat{\gamma}_i \left\langle \boldsymbol{x} - \boldsymbol{z}, \boldsymbol{v}\right\rangle^2\right). \quad (5.156)$$

Taking advantage of the definition of $\mathfrak{D}_0^n$ we conclude that there exists $c > 0$ such that

$$\frac{\mathcal{F}[\phi^0]\left(\boldsymbol{x}, \boldsymbol{x} - \hat{\boldsymbol{e}}^{k,r} + \hat{\boldsymbol{e}}^{k,s}\right)}{\mathcal{Q}^n(\boldsymbol{z})\, r_{r,s}^k\left(v_r^k - v_s^k\right)} \leq \exp\left(-c\, N^{2\delta}\right) \quad (5.157)$$

uniformly for all $\boldsymbol{x} \in \mathfrak{D}^n \setminus \mathfrak{D}_0^n$, $k = 1, \ldots, n$ and $(r,s) \in \Delta_k$. Hence, by choosing the constant, $c(N)$, in the definition of $\mathfrak{f}$, see (5.139), appropriately, (5.149) follows.

(ii) Notice that a computation along the lines of the proof of Lemma 5.20 reveals that there exists $c_{20} > 0$ such that

$$\frac{\left|\mathrm{d}\mathcal{F}[\phi^0](\boldsymbol{x})\right|}{\mathcal{F}[\phi^0]\left(\boldsymbol{x}, \boldsymbol{x} - \hat{\boldsymbol{e}}^{k,r} + \hat{\boldsymbol{e}}^{k,s}\right)} = \left|\left\langle \boldsymbol{v}, R\left(A^n(\boldsymbol{z}) + |\hat{\gamma}_1|\boldsymbol{v} \cdot \boldsymbol{v}^{\mathrm{T}}\right)(\boldsymbol{x} - \boldsymbol{z})\right\rangle + \mathcal{O}(\delta_N^2)\right|$$

$$\leq c_{20}\, \delta_N^2 \quad (5.158)$$

uniformly for $\boldsymbol{x} \in \mathfrak{D}^n$, $k = 1, \ldots, n$ and $(r,s) \in \Delta_k$. Here, the second step relies on (5.108). In order to establish (5.150) and (5.151) we prove recursively a bound on the correction u. To start with, let d_j be the smallest constants such that for all $\boldsymbol{x} \in \mathfrak{L}_j$

$$\left|\mathcal{F}[u]\left(\boldsymbol{x}, \boldsymbol{x} - \hat{\boldsymbol{e}}^{k,r} + \hat{\boldsymbol{e}}^{k,s}\right)\right| \leq d_j\, \delta_N^2\, \mathcal{F}[\phi^0]\left(\boldsymbol{x}, \boldsymbol{x} - \hat{\boldsymbol{e}}^{k,r} + \hat{\boldsymbol{e}}^{k,s}\right), \quad (5.159)$$

for all $k = 1, \ldots, n$ and $(r,s) \in \Delta_k$. Then, for any $\boldsymbol{x} \in \mathfrak{L}_{j+1} \cap \mathfrak{D}_1^n$, the construction of u given in (5.147) combined with (5.158) implies

$$\frac{\left|\mathcal{F}[u]\left(\boldsymbol{x}, \boldsymbol{x} - \hat{\boldsymbol{e}}^{k,r} + \hat{\boldsymbol{e}}^{k,s}\right)\right|}{\mathcal{F}[\phi^0]\left(\boldsymbol{x}, \boldsymbol{x} - \hat{\boldsymbol{e}}^{k,r} + \hat{\boldsymbol{e}}^{k,s}\right)}$$

$$\leq q_{r,s}^k \left(\sum_{l=1}^n \sum_{(r',s') \in \Delta_k} \frac{\left|\mathcal{F}[u]\left(\boldsymbol{x} + \hat{\boldsymbol{e}}^{l,r'} - \hat{\boldsymbol{e}}^{l,s'}, \boldsymbol{x}\right)\right|}{\mathcal{F}[\phi^0]\left(\boldsymbol{x}, \boldsymbol{x} - \hat{\boldsymbol{e}}^{k,r} + \hat{\boldsymbol{e}}^{k,s}\right)} + c_{20}\, \delta_N^2\right)$$

$$\leq \delta_N^2 \left(d_j\, q_{r,s}^k \sum_{l=1}^n \sum_{(r',s') \in \Delta_k} \frac{\mathcal{F}[\phi^0]\left(\boldsymbol{x} + \hat{\boldsymbol{e}}^{l,r} - \hat{\boldsymbol{e}}^{l,s}, \boldsymbol{x}\right)}{\mathcal{F}[\phi^0]\left(\boldsymbol{x}, \boldsymbol{x} - \hat{\boldsymbol{e}}^{k,r'} + \hat{\boldsymbol{e}}^{k,s'}\right)} + c_{20}\right).$$

5.4. Lower bounds on capacities

By plugging in the definition of ϕ^0, we have that

$$q_{r,s}^k \sum_{l=1}^n \sum_{(r',s') \in \Delta_k} \frac{\mathcal{F}[\phi^0](x + \hat{e}^{l,r} - \hat{e}^{l,s}, x)}{\mathcal{F}[\phi^0](x, x - \hat{e}^{k,r'} + \hat{e}^{k,s'})}$$

$$= 1 + q_{r,s}^k \frac{\langle v, R(A^n(z) + |\hat{\gamma}_1| v \cdot v^T)(x-z) \rangle}{r_{r,s}^k(v_r^k - v_s^k)} + \mathcal{O}(\delta_N^2)$$

$$= 1 + \mathcal{O}(\delta_N^2) \tag{5.160}$$

uniformly for all $x \in \mathfrak{D}_1^n$, $k = 1, \ldots, n$ and $(r,s) \in \Delta_k$. Here, the first step holds true due to the choice of the probability distribution, $q_{r,s}^k$, while in the second step we exploit again (5.108). Thus, the constants d_j satisfy the following recursive bound

$$d_0 = 0, \qquad d_{j+1} \leq d_j(1 + \mathcal{O}(\delta_N^2)) + c_{20}. \tag{5.161}$$

Since $T \leq c_{19}(n) N \delta_N$, we obtain from (5.161) that for all $1 \leq j \leq T$ and N sufficiently large

$$d_j \leq j\, c_{20}\, e^{j\,\mathcal{O}(\delta_N^2)} \leq T c_{20}\, e^{T\,\mathcal{O}(\delta_N^2)} \leq \mathcal{O}(1/\delta_N). \tag{5.162}$$

As a result, the function u satisfies

$$|\mathcal{F}[u](x, x - \hat{e}^{k,r} + \hat{e}^{k,s})| \leq \mathcal{O}(\delta_N^2) \mathcal{F}[\phi^0](x, x - \hat{e}^{k,r} + \hat{e}^{k,s}) \tag{5.163}$$

uniformly for all $x \in \mathfrak{D}_1^n$, $k = 1, \ldots, n$ and $(r,s) \in \Delta_k$. In particular, together with (5.155) this concludes the proof of (5.150).

(iii) In view of (5.163), (5.157) and (5.115), (5.151) is satisfied uniformly on $\mathfrak{D}_1^n \setminus \mathfrak{D}_0^n$. Moreover, due to our construction of the flow, $\phi \equiv 0$ on $\mathfrak{D}^n \setminus \mathfrak{D}_2^n$ and (5.151) is immediately satisfied in the later domain. Thus, it remains to control $\mathfrak{f}$ on $\mathfrak{D}_2^n \setminus \mathfrak{D}_1^n$. Since $\phi(x) = u(x)$ and $\mathrm{d}\mathcal{F}[u](x) = 0$ for all $x \in \mathfrak{D}_2^n \setminus \mathfrak{D}_1^n$, in the worst case, the flow $\mathfrak{f}(x, x - \hat{e}^{k,r} + \hat{e}^{k,s})$ could be equal to the total flow through $\mathfrak{D}_2^n \setminus \mathfrak{D}_1^n$ that is

$$\sum_{x \in \mathfrak{D}_1^n} \sum_{k=1}^n \sum_{(r,s) \in \Delta_k} \mathfrak{f}(x, x - \hat{e}^{k,r} + \hat{e}^{k,s})\, \mathbb{1}_{x - \hat{e}^{k,r} + \hat{e}^{k,s} \in \mathfrak{D}_2^n}. \tag{5.164}$$

However, due to (5.163) and (5.157), the latter is of order $\mathcal{O}(N^{n/2} e^{-cN^{2\delta}})$. Hence, by combining the estimates on the different domains, we conclude that there exists $\kappa > 0$ such that (5.151) holds. □

Corollary 5.26. *Let $\mathfrak{f}$ be the flow constructed above. Then, for the associated Markov chain with the law $\mathbb{P}^{\mathfrak{f}}$ it holds that*

$$\mathbb{P}^{\mathfrak{f}}\left[\sum_{(x,y) \in \gamma} \phi(x,y) = 1 + o_N(1)\right] = 1 - o_N(1). \tag{5.165}$$

Proof. By (5.149), $\mathfrak{f}$ is a unit flow that is non-negative by construction. Hence, we can associate a Markov chain $\boldsymbol{X} = \{\boldsymbol{X}(t)\}$ to it. For some $0 < \varepsilon \ll 1$, we denote by $\partial_A \mathfrak{D}_{0,\varepsilon}^n \equiv \partial_A \mathfrak{D}_0^n(z, \varepsilon \nu_0 \delta_N)$ the set of all configurations at the boundary $\partial_A \mathfrak{D}_0^n$ that lies deeply inside the narrow tube. Now, let us consider the set of all realizations of $\boldsymbol{X}$ from $\partial_A \mathfrak{D}_{0,\varepsilon}^n$ to $\partial_B \mathfrak{D}_0^n$ that stays inside $\mathfrak{D}_0^n$, i.e.

$$\mathcal{P}_{A,B} := \{\boldsymbol{x} \mid \boldsymbol{x}(a_{\boldsymbol{x}}) \in \partial_A \mathfrak{D}_{0,\varepsilon}^n, \, \boldsymbol{x}(b_{\boldsymbol{x}}) \in \partial_B \mathfrak{D}_0^n, \, \boldsymbol{x}(t) \in \mathfrak{D}_0^n, \, \forall \, a_{\boldsymbol{x}} < t < b_{\boldsymbol{x}}\}.$$

In view of (5.150), we have for all $\gamma \in \mathcal{P}_{A,B}$ that

$$\sum_{(\boldsymbol{x},\boldsymbol{y}) \in \gamma} \phi(\boldsymbol{x}, \boldsymbol{y}) = \big(g(a_{\boldsymbol{x}}) - g(b_{\boldsymbol{x}})\big) (1 + o_N(1)) = 1 + o_N(1), \quad (5.166)$$

where we used that the function g, as defined in (5.97), converge exponentially fast in N to one for all $\boldsymbol{x} \in \partial_A \mathfrak{D}^n$ and to zero for all $\boldsymbol{x} \in \partial_B \mathfrak{D}^n$.

Hence, it remains to show that the $\mathbb{P}^{\mathfrak{f}}$-probability of the event $\mathcal{P}_{A,B}$ is close to one. Recall that the initial distribution of the Markov chain $\boldsymbol{X}$ is given by $\sum_{\boldsymbol{y} \in \mathfrak{D}^n} \mathfrak{f}(\boldsymbol{x}, \boldsymbol{y})$ for all $\boldsymbol{x} \in \partial_A \mathfrak{D}^n$. Its transition probabilities are given by $q^{\mathfrak{f}}(\boldsymbol{x}, \boldsymbol{y}) = \mathfrak{f}(\boldsymbol{x}, \boldsymbol{y}) / \sum_{\boldsymbol{y}'} \mathfrak{f}(\boldsymbol{x}, \boldsymbol{y}')$. Since the flow $\mathfrak{f}$ is defined in terms of ϕ^0 on $\partial_A \mathfrak{D}_0^n$, (5.156) implies that the set of all path starting in $\partial_A \mathfrak{D}^n \setminus \partial_A \mathfrak{D}_{0,\varepsilon}^n$ is negligible with respect to $\mathbb{P}^{\mathfrak{f}}$. Now, set $\boldsymbol{Y}(t) := \boldsymbol{X}(t) - \check{\boldsymbol{v}} \langle \boldsymbol{X}(t), \check{\boldsymbol{v}} \rangle / \|\check{\boldsymbol{v}}\|_2^2$ and decompose $\boldsymbol{Y}(t) = \boldsymbol{Y}(0) + \boldsymbol{M}(t) + \boldsymbol{N}(t)$ into a martingale $\boldsymbol{M}(t)$ and a previsible process $\boldsymbol{N}(t)$. As a consequence of (5.163), for all $t \leq T$, $\|\boldsymbol{N}(t)\| \leq T/N \mathcal{O}(\delta_N) = \delta_N \mathcal{O}(\delta_N)$ on the event $\{\boldsymbol{X}(t) \in \mathfrak{D}_1^n, \, \forall \, 0 \leq t \leq T\}$. Hence, Doob's maximums inequality for submartingales implies

$$\mathbb{P}\left[\max_{0 \leq t \leq T} \|\boldsymbol{Y}(t)\| < \frac{1}{2} \nu_0 \delta_N\right] \geq 1 - \mathbb{P}\left[\max_{0 \leq t \leq T} \|\boldsymbol{M}(t)\|^2 \geq c^2 \delta_N^2\right]$$

$$\geq 1 - \frac{1}{c^2 \delta_N^{-2}} \mathbb{E}\big[\|\boldsymbol{M}(T)\|^2\big]$$

$$\geq 1 - \mathcal{O}(N^{-1/2-\delta}), \quad (5.167)$$

where $c < \nu_0 (1/2 - \varepsilon) - \mathcal{O}(\delta_N)$ and $\mathbb{E}\big[\|\boldsymbol{M}(T)\|^2\big] \leq T \mathcal{O}(N^{-2})$. This completes the proof. $\square$

The remaining part of this subsection is devoted to the construction of the flow $\mathfrak{f}_A$ from $\boldsymbol{A}$ to $\partial_A \mathfrak{D}^n$ of the form

$$\mathfrak{f}_A(\boldsymbol{x}, \boldsymbol{y}) = \frac{\mathcal{Q}^n(\boldsymbol{x}) \, r^n(\boldsymbol{x}, \boldsymbol{y})}{\mathcal{E}^n(\tilde{g})} \phi_A(\boldsymbol{x}, \boldsymbol{y}) \, (1 + o_N(1)) \quad (5.168)$$

and a of the corresponding flow $\mathfrak{f}_B$ from $\partial_B \mathfrak{D}^n$ to $\boldsymbol{B}$ such that the concatenation $\mathfrak{f}_{A,B} = \{\mathfrak{f}_A, \mathfrak{f}, \mathfrak{f}_B\}$ satisfies Kirchhoffs law and

$$\mathbb{P}^{\mathfrak{f}_A}\left[\sum_{(\boldsymbol{x},\boldsymbol{y}) \in \gamma} \phi_A(\boldsymbol{x}, \boldsymbol{y}) = o_N(1)\right] = 1 - o_N(1) \quad (5.169)$$

5.4. Lower bounds on capacities

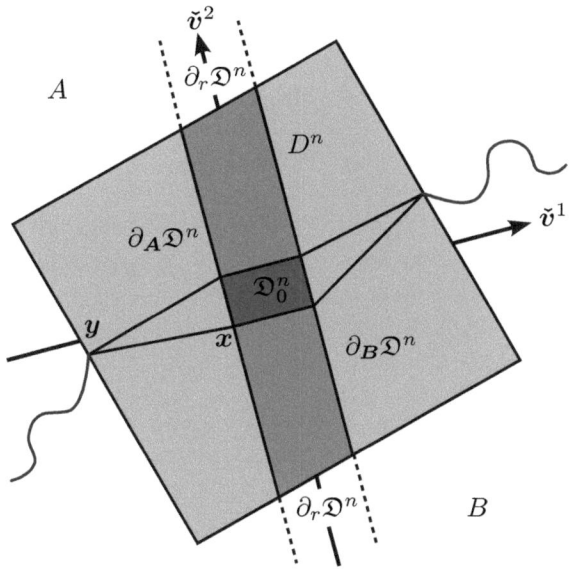

Figure 5. Construction of the paths $\hat{\gamma}^x = \{\hat{\gamma}, \hat{\eta}^x\}$ for each $x \in \partial_A \mathfrak{D}_0^n$. While $\hat{\gamma}$ is defined in terms of the minimal energy curve that ends in some point $y \in D^n$, $\hat{\eta}^x$ is just the linear interpolation between y and x.

and

$$\mathbb{P}^{\mathfrak{f}_B}\left[\sum_{(x,y)\in\gamma} \phi_B(x,y) = \mathcal{O}_N(1)\right] = 1 - o_N(1). \tag{5.170}$$

Together with (5.165) these estimates imply that (5.127) holds true. Since the construction of $\mathfrak{f}_A$ and $\mathfrak{f}_B$ are completely similar, we will work out the details only for $\mathfrak{f}_A$.

The strategy is the following. To each $x \in \partial_A \mathfrak{D}^n$ associate a nearest neighbor path $\hat{\gamma}^x = (\hat{\gamma}^x(a^x), \ldots, \hat{\gamma}^x(0))$ on Γ^n that starts in A and ends in x, i.e.

$$\hat{\gamma}^x(a^x) \in A, \quad \hat{\gamma}^x(0) = x \quad \text{and} \quad \|\hat{\gamma}^x(t+1) - \hat{\gamma}^x(t)\|_1 = \frac{2}{N},$$

for all $t = a^x, \ldots, -1$. Then, the flow from A to $\partial_A \mathfrak{D}^n$ is given by

$$\mathfrak{f}_A(y, y') := \sum_{x \in \partial_A \mathfrak{D}_0^n} \mathbb{1}_{(y,y')\in\hat{\gamma}^x} \sum_{k=1}^n \sum_{(r,s)\in\Delta_k(x)} \mathfrak{f}(x, x - \hat{e}^{k,r} + \hat{e}^{k,s}).$$

Notice that the flow $\mathfrak{f}_A$ constructed above as well as its concatenation $\{\mathfrak{f}_A, \mathfrak{f}\}$ with the flow $\mathfrak{f}$ satisfies Kirchhoff's law. The idea, we have in mind to satisfy (5.169), is to choose the paths $\hat{\gamma}^x$ is such a way that $F^n(z) - F^n(\hat{\gamma}^x(t)) > \varepsilon \delta_N^2$ for all $t = a^x, \ldots, 0$ for some $\varepsilon > 0$ small enough.

In the sequel, we describe the actual construction of the family of paths $\{\hat{\gamma}^x\}$ for $x \in \partial_A \mathfrak{D}^n$. Fix a $x \in \partial_A \mathfrak{D}^n$. The path $\hat{\gamma}^x$ associated to x consists of a concatenation of two paths $\hat{\gamma}^x = \{\hat{\gamma}, \hat{\eta}^x\}$ where $\hat{\gamma}$ connects A with some point $y \in \partial_A D^n$ and $\hat{\eta}^x$ runs from y to $x \in \partial_A \mathfrak{D}^n$, see Figure 5.4.2.

Let $m \equiv m(\boldsymbol{m}) \in \mathcal{M}_1$ and $m' \equiv m'(\boldsymbol{m}') \in \mathcal{M}_1$ be local minima of the function F_N that correspond to the minima $\boldsymbol{m}, \boldsymbol{m}'$ of F^n. In view of Lemma 5.8 it holds that $m(\boldsymbol{m}) = \sum_{k=1}^n \boldsymbol{m}^k$ and likewise for m'. In order to construct $\hat{\gamma}$, consider the minimal energy curve $\varphi : [0,1] \to \mathcal{M}_1$ with respect to F_N that connects the minimum m and m'. In particular, this curve passes through the saddle point $z \equiv z(\boldsymbol{z})$. Further, the mesoscopic path $\boldsymbol{\varphi}$ corresponding to φ is uniquely defined by (5.54) and (5.55). We denote its nearest neighbor path approximation in Γ^n by $\hat{\varphi}$. Let $\boldsymbol{a} \in \boldsymbol{A}$ be the last point before $\hat{\varphi}$ leaves the set $\boldsymbol{A}$ and $\boldsymbol{y} \in \boldsymbol{D}^n$ be the first point where $\hat{\varphi}$ enters the set $\boldsymbol{D}^n$. Then, the path $\hat{\gamma}$ is given by the portion of the path $\hat{\varphi}$ from $\boldsymbol{a}$ to $\boldsymbol{y}$. On the other hand, for any $\boldsymbol{x} \in \partial_A \mathfrak{D}^n$, let $\boldsymbol{\eta}^x : [0,1] \to \times_{k=1}^n \mathcal{M}_{\pi_k}, t \mapsto \boldsymbol{y} + t(\boldsymbol{x} - \boldsymbol{y})$ be the line segment that connects $\boldsymbol{y}$ and $\boldsymbol{x}$. Then, $\hat{\eta}^x$ is defined as the nearest neighbor approximation of $\boldsymbol{\eta}^x$ in Γ^n.

Lemma 5.27. *Let $\mathfrak{f}_A$ be the flow constructed above. Then, for the associated Markov chain with law $\mathbb{P}^{\mathfrak{f}_A}$, (5.169) is satisfied.*

Proof. First of all note that, in view of (5.151), the contribution to the flow $\mathfrak{f}_A$ coming from paths $\hat{\gamma}^x$ with $x \in \partial_A \mathfrak{D}^n \setminus \partial_A \mathfrak{D}_0^n$ is negligible. By plugging the expressions for $\mathcal{Q}^n(x)$, $r^n(x, y)$ and $\mathcal{E}^n(\tilde{g})$ given by (5.45), (5.79) and (5.115) into the right-hand side of (5.168), we obtain immediately an upper bound for $\phi_A(x, y)$. Namely,

$$\phi_A(x, y) \leq 2 \frac{\mathcal{E}^n(\tilde{g}) \, \mathfrak{f}_A(x,y)}{\mathcal{Q}^n(x) \, r^n(x,y)} \leq c N^{n(q-1)/2} \exp\!\left(-\beta N \bigl(F^n(z) - F^n(x)\bigr)\right). \tag{5.171}$$

Thus, it remains to show that there exists $\varepsilon > 0$ such that $F^n(z) - F^n(\hat{\gamma}(t)) > \varepsilon \delta_N^2$ for all $t = a^x, \ldots, 0$.

As a consequence of the fact that F_N is essential smooth with respect to $\mathcal{M}_1$, there exists a $\delta > 0$ such that distance between the minimal energy curve and the boundary of $\mathcal{M}_1$ is at least δ. Thus, by construction, the minimal distance between φ and the boundary of $\times_{k=1}^n \mathcal{M}_{\pi_k}$ is at least $\delta \min_k \pi_k$. In particular, $\|\nabla F_N(\varphi(t))\|$ is for all $t \in [0,1]$ uniformly bounded from above by a constant that is independent of N. Recall that z is the unique minimal saddle point between the sets A and B. By exploiting the boundedness of $\|\nabla F^n(\varphi(t))\|$ and the fact that the Hausdorff distance between φ and $\hat{\gamma}$ is at most $\sqrt{n(q-1)/q}/N$, we conclude that $\max_{x \in \hat{\gamma}} F^n(x) \leq F^n(y) + \mathcal{O}(n/N)$.

5.4. Lower bounds on capacities

Moreover, assuming that the parameter ν in the definition of $\mathfrak{D}^n(z, \nu\delta_N)$ is so small that $\mathfrak{D}^n$ lies deeply inside the set $D^n(z, \delta_N)$, we can ensure that

$$F^n(y) < \min_{x \in \partial_A \mathfrak{D}_0^n} F^n(x). \tag{5.172}$$

Further, by choosing ν_0 small enough, we can ensure that $\langle A^n(z)(x - y), x - z\rangle > 0$ uniformly for all $x \in \partial_A \mathfrak{D}_0^n$. Hence, for all $t \in [0, 1]$

$$\langle A^n(z)(y - z + t(x - y)), (y - z + t(x - y))\rangle \leq \langle A^n(z)(x - z), x - z\rangle,$$

i.e. up to negligible corrections $F^n(y + t(x - y)) \leq F^n(x)$ uniformly for all $t \in [0, 1]$ and $x \in \partial_A \mathfrak{D}_0^n$. Hence, $\max_{x' \in \hat{\gamma}^x} F^n(x') \leq F^n(x) + \mathcal{O}(n/N)$ and there exists a $\varepsilon > 0$ such that $F^n(z) - F^n(x) > \varepsilon \delta_N$ uniformly for all $x \in \partial_A \mathfrak{D}_0^n$. □

5.4.3. Statistics of mesoscopic trajectories. In order to construct a subordinate flow it suffices to consider a certain subset of all realizations of the mesoscopic chain $\{X(t)\}$ that starts in A and is stopped at the arrival of B. Recall our convention to choose the label of a realization $\boldsymbol{x} = (x(a_{\boldsymbol{x}}), \ldots, x(0), \ldots, x(b_{\boldsymbol{x}}))$ such that $x(a_{\boldsymbol{x}}) \in A$, $x(b_{\boldsymbol{x}}) \in B$ and $\sum_{k=1}^n x(0)^k = \sum_{k=1}^n z^k$. For a given path $\boldsymbol{x}$, we denote by $\tau_k[T] \equiv \tau_k[T](\boldsymbol{x})$ the number of changed in the k's block compared to $x(0)$ along the path $\boldsymbol{x}$ until time T, i.e. for $T > 0$ let

$$\tau_k[T] = \frac{N}{2} \sum_{t=0}^{T-1} \|x^k(t) - x^k(t-1)\|_1$$

while for $T < 0$ the sum runs from $T + 1$ to 0.

The set of paths, we are focusing on in the construction of a subordinate flow, are characterized by the following

Definition 5.28. A mesoscopic path $\boldsymbol{x} = (x(a_{\boldsymbol{x}}), \ldots, x(b_{\boldsymbol{x}}))$ is called *good* if the following properties are satisfied:

(i) $\boldsymbol{x}$ passes through $\mathfrak{D}_0^n$.
(ii) there exists $C_1 > 0$ such that

$$\max_{k=1,\ldots,n} \frac{\tau_k[b_{\boldsymbol{x}}]}{|\Lambda_k|} \leq C_1 \quad \text{and} \quad \max_{k=1,\ldots,n} \frac{\tau_k[a_{\boldsymbol{x}}]}{|\Lambda_k|} \leq C_1. \tag{5.173}$$

(iii) it holds that

$$\sum_{t=a_{\boldsymbol{x}}}^{b_{\boldsymbol{x}}-1} \exp\left(\mathcal{O}(\varepsilon(n)) \sum_{k=1}^n \frac{\tau_k[t]^2}{|\Lambda_k|}\right) \phi_{A,B}(x(t), x(t+1)) \leq 1 + \mathcal{O}(\varepsilon(n)). \tag{5.174}$$

The set of all good paths is denoted by $\mathcal{GP}_{A,B}$.

Lemma 5.29 ([6, Proposition 5.4]). *Let $\mathfrak{f}_{A,B}$ be the mesoscopic flow constructed in the previous subsection and denote by $\boldsymbol{X} = \{\boldsymbol{X}(t)\}_t$ the Markov chain on Γ^n with law $\mathbb{P}^{\mathfrak{f}_{A,B}}$. Then,*

$$\mathbb{P}^{\mathfrak{f}_{A,B}}\left[\boldsymbol{X} \in \mathcal{GP}_{A,B}\right] = 1 - o_N(1). \tag{5.175}$$

Proof. Due to the construction of the flow $\mathfrak{f}$, the probability of the event that the associated Markov chain does not pass through $\mathfrak{D}_0^n$ is of order $o_N(1)$. As a further consequence, the step frequencies, $\tau_k[t]/t$, are on average proportional to π_k. Therefore, there exist a constant C_1 such that, up to an exponentially small $\mathbb{P}^{\mathfrak{f}_{A,B}}$-probability, it holds that

$$\max_{k=1,\ldots,n} \frac{\tau_k[b_{\boldsymbol{x}}]}{|\Lambda_k|} \leq C_1.$$

Hence, it remains to show that the $\mathbb{P}^{\mathfrak{f}_{A,B}}$-probability of the event (5.174) is close to one. In order to do so, notice that from the construction of the mesoscopic flow, namely the fact that the transition probabilities $q^{\mathfrak{f}}$ inside $\mathfrak{D}_0^n$ are given by (5.148) up to small correction, and from the property (5.57) of the minimal energy curve φ, it follows that there exists a $K < \infty$, independent of n, such that

$$\mathbb{P}^{\mathfrak{f}_{A,B}}\left[\max_{k=1,\ldots,n} \max_{|t| > N^{1/2-\delta}} \frac{\tau_k[t]}{|t|\pi_k} > K\right] = o_N(1). \tag{5.176}$$

Now, observe that uniformly for all $t < N^{1/2-\delta}$ and all paths $\boldsymbol{x}$ it holds that

$$\sum_{k=1}^{n} \frac{\tau_k[t]^2}{|\Lambda_k|} = \mathcal{O}(N^{-2\delta}). \tag{5.177}$$

On the other hand, for all paths $\boldsymbol{x}$ that satisfy the condition $\tau_k[t] \leq K\pi_k|t|$ uniformly for all $|t| > N^{1/2-\delta}$ and $k = 1,\ldots,n$, we have

$$\mathcal{O}(\varepsilon(n)) \sum_{k=1}^{n} \frac{\tau_k[t]^2}{|\Lambda_k|} \leq \mathcal{O}(\varepsilon(n)) K^2 \frac{t^2}{N} \sum_{k=1}^{n} \pi_k = \mathcal{O}(\varepsilon(n)) K^2 \frac{t^2}{N}. \tag{5.178}$$

Moreover, due to the construction of $\mathfrak{f}_{A,B}$, there exists a $c > 0$ such that all those paths $\boldsymbol{x}$ satisfy $\phi_{A,B}(\boldsymbol{x}(t), \boldsymbol{x}(t+1)) \leq e^{-ct^2/2}$ for all $|t| > N^{1/2-\delta}$. Hence, in view of (5.126) we obtain that with $\mathbb{P}^{\mathfrak{f}_{A,B}}$ close to one

$$\sum_{t=a_{\boldsymbol{x}}}^{b_{\boldsymbol{x}}-1} \exp\left(\mathcal{O}(\varepsilon(n)) \sum_{k=1}^{n} \frac{\tau_k[t]^2}{|\Lambda_k|}\right) \phi_{A,B}(\boldsymbol{x}(t), \boldsymbol{x}(t+1)) = 1 + \mathcal{O}(\varepsilon(n)). \tag{5.179}$$

This yields the assertion. □

5.4. Lower bounds on capacities

5.4.4. Construction of a subordinate microscopic flow. Fix a mesoscopic path $x \in \mathcal{GP}_{A,B}$. Now, our strategy is to construct a directed Markov chain $\Sigma^x = \{\Sigma^x(t)\}_t$ on the set of configurations given by the mesoscopic path, i.e. $\mathcal{S}^n[x(a_x)] \cup \ldots \cup \mathcal{S}^n[x(b_x)]$. In view of (1.81), the subordinate flow, f^x, is then defined by $f^x(\sigma, \eta) = \mathbb{P}^{f^x}[(\sigma, \eta) \in \Sigma^x]$. Actually, since the construction of f^x from $\mathcal{S}^n[x(a_x)]$ to $\mathcal{S}^n[x(0)]$ and $\mathcal{S}^n[x(0)]$ to $\mathcal{S}^n[x(b_x)]$ is completely similar, we will work out the details only for the latter one.

To start with, let us denote by

$$\Theta_{k,t}^{\pm} := \left\{ r \in \mathcal{S}_0 \,\Big|\, x_r^k(t+1) - x_r^k(t) = \pm \tfrac{1}{N} \right\} \tag{5.180}$$

the corresponding color that is created, respectively, annihilated in the k's block from $x(t)$ to $x(t+1)$. Due to the underlying single site dynamic, for each t there exists a unique $k \in \{1, \ldots, n\}$ such that the sets $\Theta_{k,t}^+$ and $\Theta_{k,t}^-$, respectively, consists of exactly one element while for all $l \neq k$ the sets $\Theta_{l,t}^{\pm}$ are empty. For a given configuration σ, recall that we defined by

$$\Lambda_{k,r}(\sigma) := \{i \in \Lambda_k \mid \sigma_i = r\}, \qquad r \in \mathcal{S}_0$$

the set of lattice site in Λ_k those corresponding spin variables are equal to the color r.

Let us now define a time-inhomogeneous Markov chain $\{\Sigma^x(t)\}$. Its initial distribution, ν_0^x on $\mathcal{S}^n[x(0)]$ is given by the Gibbs measure, μ_N, conditioned on the set $\mathcal{S}^n[x(0)]$, and we choose the following transition probabilities

$$q_t\left(\sigma, \sigma^{i,s}\right) := \begin{cases} \dfrac{\exp\!\left(\beta \langle \tilde{h}^i, -e^r + e^s \rangle\right)}{\sum_{j \in \Lambda_{k,r}(\sigma)} \exp\!\left(\beta \langle \tilde{h}^j, -e^r + e^s \rangle\right)}, & \text{if } \Theta_{k,t}^- = r,\ \Theta_{k,t}^+ = s,\ i \in \Lambda_{k,r}(\sigma), \\ 0, & \text{otherwise.} \end{cases} \tag{5.181}$$

Thus, the subordinate flow, f^x, through the edge $(\sigma(t), \sigma(t+1))$ where $\sigma(t) \in \mathcal{S}^n[x(t)]$ and $\sigma(t+1) \in \mathcal{S}^n[\sigma(t+1)]$ is equal to

$$f^x\big(\sigma(t), \sigma(t+1)\big) = \nu_t^x\big(\sigma(t)\big)\, q_t\big(\sigma(t), \sigma(t+1)\big), \tag{5.182}$$

where ν_t^x denotes the marginal distribution of the chain.

Remark 5.30. Note that any choice of the transition probabilities q_t give rise to a subordinate flow. But, in general, the marginal distributions, ν_t^x, of this chain differ from the conditional distribution $\mu_{\mathcal{S}^n[x(t)]} \equiv \mu_N[\,\cdot\, |\mathcal{S}^n[x(t)]]$. Thus, one would like to choose the transition probabilities in such a way that $\nu_t^x = \mu_{\mathcal{S}^n[x(t)]}$ for all $t = 0, \ldots, b_x$. Indeed, in the case where all $\tilde{h}^i \equiv 0$, by choosing

$$q_t\left(\sigma, \sigma^{i,s}\right) = \begin{cases} 1/|\Lambda_{k,r}(\sigma)|, & \text{if } \Theta_{k,t}^- = r,\ \Theta_{k,t}^+ = s,\ i \in \Lambda_{k,r}(\sigma), \\ 0, & \text{otherwise,} \end{cases}$$

the transported measure ν_t^x equals $\mu_{\mathcal{S}^n[x(t)]}$ that is the uniform measure on $\mathcal{S}^n[x(t)]$. However, in the general case it is not known under which conditions on size of the distortion of $\tilde{h}^i$ a transition matrix, q_t, exists satisfying such property.

Since,

$$q_t\big(\sigma(t),\sigma(t+1)\big) = \frac{1+\mathcal{O}(\varepsilon(n))}{|\Lambda_{\Theta_{k,t}^-}(\sigma(t))|}, \qquad \frac{p_N\big(\sigma(t),\sigma(t+1)\big)}{r^n\big(x(t),x(t+1)\big)} = \frac{1+\mathcal{O}(\varepsilon(n))}{|\Lambda_{\Theta_{k,t}^-}(\sigma(t))|} \tag{5.183}$$

the expression (5.135) can be rewritten as

$$\Phi_t^x\big(\sigma(t)\big) = \frac{\nu_t^x\big(\sigma(t)\big)}{\mu_N\big[\sigma(t)\,\big|\,\mathcal{S}^n[x(t)]\big]}\big(1+\mathcal{O}(\varepsilon(n))\big) =: \Psi_t\big(\sigma(t)\big)\big(1+\mathcal{O}(\varepsilon(n))\big)$$

where Ψ_t can be seen as a measure for the deviation of ν_t^x from the measure $\mu_{\mathcal{S}^n[x(t)]}$. As a consequence of the definition of the Markov chain $\{\Sigma^x(t)\}$, $\Psi_0(\sigma) \equiv 1$ for all $\sigma \in \mathcal{S}^n[x(0)]$. However, for any $t > 0$ this is no longer true. Hence, it remains to study the average error propagation along the mesoscopic path x. In particular, our goal is to show that for any good mesoscopic path $x \in \mathcal{GP}_{A,B}$ it holds that $\mathbb{E}^{f^x}[\Psi_t] \leq \exp\big(\mathcal{O}(\varepsilon(n))\,t^2/N\big)$.

Notice that the Gibbs measure, μ_N, conditioned of $\mathcal{S}^n[x(t)]$ is a product measure,

$$\mu_N\big[\,\cdot\,\big|\,\mathcal{S}^n[x(t)]\big] = \bigotimes_{k=1}^n \mu_t^k(\,\cdot\,), \tag{5.184}$$

where μ_t^k is the canonical measure on $\mathcal{S}_{|\Lambda_k|} := \mathcal{S}_0^{\Lambda_k}$ that is given by

$$\mu_t^k(\sigma) := \frac{1}{Z[x^k(t)]}\exp\Big(\beta\sum_{i\in\Lambda_k}\langle \tilde{h}^i, e^{\sigma_i}\rangle\Big)\mathbb{1}_{\varrho^k(\sigma)=x^k(t)}, \qquad \sigma \in \mathcal{S}_0^{\Lambda_k}. \tag{5.185}$$

On the other hand, according to the choice of the transition probabilities (5.181), the Markov chain $\{\Sigma^x(t)\}$ splits into a direct product of n Markov chains, denoted by $\{\Sigma^{x,1}(t)\},\ldots,\{\Sigma^{x,n}(t)\}$, on the state spaces $\mathcal{S}_{|\Lambda_1|},\ldots,\mathcal{S}_{|\Lambda_n|}$ that evolves independently. Hence, if the Markov chain $\{\Sigma^x(t)\}$ has performed T steps, each of the chains $\{\Sigma^{x,k}(t)\}$ performs exactly $\tau_k[T]$ steps. Thus, the corrector, Ψ_t, is equal to

$$\Psi_t\big(\sigma(t)\big) = \prod_{k=1}^n \psi_{\tau_k[t]}^k\big(\sigma_{\Lambda_k}(t)\big), \tag{5.186}$$

where $\sigma_{\Lambda_k}(t)$ is the projection of $\sigma(t)$ onto $\mathcal{S}_{|\Lambda_k|}$. Therefore, it suffices to study the propagation of errors with respect to each $\{\Sigma^{x,k}(t)\}$.

Proposition 5.31. *Fix a $k \in \{1,\ldots,n\}$ and let us denote by $\mathbb{P}_{\mu_0^k}$ the law of the Markov chain $\{\Sigma^{x,k}(t)\}$ on $\mathcal{S}_{|\Lambda_k|}$. Then, there exists $c < \infty$ such that*

$$\mathbb{E}_{\mu_0^k}[\psi_t^k] \leq \exp\Big(c\varepsilon(n)\,\frac{t^2}{|\Lambda_k|}\Big), \qquad \forall\, t = 0,\ldots,\tau_k[b_x]. \tag{5.187}$$

5.4. Lower bounds on capacities

In view of Proposition 5.31, we obtain the following lower bound for capacities.

Corollary 5.32. *With the notation introduced above, we obtain*

$$\mathrm{cap}(A,B) \geq \frac{\beta|\bar{\gamma}_1|}{2\pi N} \frac{\exp(-\beta N F_N(z))}{\sqrt{\det(I_q - 2\beta \mathrm{Hess}\, U_N(2\beta z))}} (1 - o_N(1)). \quad (5.188)$$

Proof of Proposition 5.31. To keep notation simple, we will use the following abbreviations

$$w(\sigma) \equiv \exp\left(\beta \sum_{i \in \Lambda_k} \langle \tilde{h}^i, e^{\sigma_i} \rangle\right), \quad \forall \sigma \in \mathcal{S}_{|\Lambda_k|}, \quad \text{and} \quad \tilde{x}(s) \equiv x^k(\tau_k^{-1}[s])$$

where $\tau_k^{-1}[s] := \min\{t \in \mathbb{N}_0 \mid \tau_k[t] \geq s\}$ and $x^k(t) \in \mathbb{R}^q$ is the k's block of $x(t)$ for a given path $\underline{x} = (x(0), \ldots, x(b_{\underline{x}}))$. Further, we define a family of functions $d_t \equiv d_t^k$: $\mathcal{S}_{|\Lambda_k|} \to \mathbb{R}$,

$$\sigma \mapsto d_t(\sigma) := \sum_{i \in \Lambda_{k,r}(\sigma)} \exp(\beta \langle \tilde{h}^i, -e^r + e^s \rangle), \quad \text{for } r = \Theta_{k,t}^-, \ s = \Theta_{k,t}^+. \quad (5.189)$$

The proof of this proposition comprises three steps.

STEP 1. In the sequel, let $\mathcal{S}[x] := \{\sigma \in \mathcal{S}_{|\Lambda_k|} \mid \varrho^k(\sigma) = x\}$ for $x \in \mathbb{R}^q$. We start with showing that for all $t = 0, \ldots, \tau_k[b_{\underline{x}}] - 1$ and $\eta \in \mathcal{S}[\tilde{x}(t)]$

$$\psi_t^k(\eta) = \frac{1}{|\mathfrak{P}_t(\eta)|} \sum_{\sigma \in \mathfrak{P}_t(\eta)} \prod_{s=0}^{t-1} \frac{\mu_s^k[d_s]}{d_s(\sigma(s))}, \quad (5.190)$$

where $\mathfrak{P}_t(\eta)$ denotes the set of all paths $\sigma = (\sigma(0), \ldots, \sigma(t-1), \eta)$ from $\mathcal{S}[\tilde{x}(0)]$ to η of positive $\mathbb{P}_{\mu_0^k}$-probability. Let us point out that an analog statement was shown in [6, Proposition 5.1]. To lighten notation, for a fixed t we set $r = \Theta_{k,t-1}^-$ and $s = \Theta_{k,t-1}^+$. Since the marginal distribution, ν_t^k, of the Markov chain $\{\Sigma^{\underline{x},k}(t)\}$ satisfies for $\eta \in \mathcal{S}[\tilde{x}(t)]$ the recursion

$$\nu_t^k(\eta) = \sum_{i \in \Lambda_{k,s}(\eta)} \nu_{t-1}^k(\eta^{i,r}) q_{t-1}(\eta^{i,r}, \eta), \quad (5.191)$$

it follows that

$$\psi_t^k(\eta) = \sum_{i \in \Lambda_{k,s}(\eta)} \frac{\nu_{t-1}^k(\eta^{i,r})}{\mu_t^k(\eta)} q_{t-1}(\eta^{i,r}, \eta) = \sum_{i \in \Lambda_{k,r}(\eta)} \frac{\mu_{t-1}^k(\eta^{i,r})}{\mu_t^k(\eta)} q_{t-1}(\eta^{i,r}, \eta) \psi_{t-1}^k(\eta^{i,r}).$$

Plugging our choice of the transition probabilities in (5.181), we obtain

$$\frac{\mu_{t-1}^k(\eta^{i,r})}{\mu_t^k(\eta)} q_{t-1}(\eta^{i,r}, \eta) = \frac{Z[\tilde{x}(t)]}{Z[\tilde{x}(t-1)]} \left(\sum_{i \in \Lambda_{k,r}(\eta^{i,r})} \exp(\beta \langle \tilde{h}^i, -e^r + e^s \rangle)\right)^{-1}.$$

As a consequence of the single site dynamics, we consider, it holds that $|\Lambda_{k,r}(\sigma)| = |\Lambda_{k,r}|$ for all $\sigma \in \mathcal{S}[\tilde{x}(t)]$. In particular, $|\Lambda_{k,r}| = \mathcal{O}(N)$. Notice that this property is implied by the fact that the minimal energy curve, $\gamma(t)$, used in the construction of the mesoscopic flow, is bounded away from the boundary of $\mathcal{M}_1$ uniformly in N. Thus, by considering the ratio of the partition functions we get

$$
\begin{aligned}
\frac{Z[\tilde{x}(t)]}{Z[\tilde{x}(t-1)]} &= \frac{1}{Z[\tilde{x}(t-1)]} \sum_{\eta \in S[\tilde{x}(t)]} \frac{1}{|\Lambda_{k,s}|} \sum_{i \in \Lambda_{k,s}(\eta)} w(\eta) \\
&= \frac{1}{Z[\tilde{x}(t-1)]} \sum_{\eta \in S[\tilde{x}(t)]} \frac{1}{|\Lambda_{k,s}|} \sum_{i \in \Lambda_{k,s}(\eta)} \exp(\beta \langle \tilde{h}^i, -e^r + e^s \rangle) w(\eta^{i,r}) \\
&= \frac{1}{Z[\tilde{x}(t-1)]} \sum_{\eta \in S[\tilde{x}(t-1)]} \frac{1}{|\Lambda_{k,s}|} \sum_{i \in \Lambda_{k,r}(\eta)} \exp(\beta \langle \tilde{h}^i, -e^r + e^s \rangle) w(\eta) \\
&= \frac{1}{|\Lambda_{k,s}|} \mu^k_{t-1}[d_{t-1}],
\end{aligned}
\qquad (5.192)
$$

where $\mu^k_t[f] = \sum_{\sigma \in S[\tilde{x}(t)]} f(\sigma) \mu^k_t(\sigma)$ denotes the expectation with respect to the measure μ^k_t. As a result,

$$
\psi^k_t(\eta) = \frac{1}{|\Lambda_{k,s}|} \sum_{i \in \Lambda_{k,s}(\eta)} \frac{\mu^k_{t-1}[d_{t-1}]}{d_{t-1}(\eta^{i,r})} \psi^k_{t-1}(\eta^{i,r}). \qquad (5.193)
$$

Iterating the argument above and using that $\psi^k_0 \equiv 1$ yields (5.190).

STEP 2. After having established a different representation for ψ^k_t, our next task is to compute the average error propagation along the path $(\tilde{x}(0), \ldots, \tilde{x}(\tau_k[b_x]))$. Let us remark that the configurations $\sigma(s)$ and $\xi(s)$ for two different microscopic paths $\sigma, \xi \in \mathfrak{P}_t(\eta)$ with $\eta \in S[\tilde{x}(t)]$ differs at most in $2(t-s)$ coordinates. Therefore, it is easy to see that there exists $c < \infty$ such that

$$
\frac{d_s(\sigma(s))}{d_s(\xi(s))} = 1 - \frac{d_s(\sigma(s)) - d_s(\xi(s))}{d_s(\xi(s))} \leq \exp\left(c\varepsilon(n) \frac{t-s}{|\Lambda_k|}\right). \qquad (5.194)
$$

Hence,

$$
\begin{aligned}
\mathbb{E}_{\mu^k_0}&[\psi^k_t] \\
&= \sum_{\eta(0),\ldots,\eta(t)} \mu^k_0(\eta(0)) \prod_{s=1}^{t} q_s(\eta(s-1), \eta(s)) \psi^k_t(\eta(t)) \\
&= \sum_{\eta(0),\ldots,\eta(t)} \mu^k_0(\eta(0)) \frac{1}{|\mathfrak{P}_t(\eta(t))|} \sum_{\sigma \in \mathfrak{P}_t(\eta(t))} \prod_{s=1}^{t} \frac{\mu^k_{s-1}[d_{s-1}]}{d_{s-1}(\sigma(s-1))} q_s(\eta(s-1), \eta(s)) \\
&\leq \sum_{\eta(0),\ldots,\eta(t-1)} \mu^k_0(\eta(0)) \frac{\mu^k_0[d_0]}{d_0(\eta(0))} \prod_{s=1}^{t-1} \frac{\mu^k_s[d_s]}{d_s(\eta(s))} q_s(\eta(s-1), \eta(s)) \exp\left(c\varepsilon(n) \frac{t^2}{|\Lambda_k|}\right).
\end{aligned}
\qquad (5.195)
$$

5.4. Lower bounds on capacities

Note that the constant c may change from line to line throughout the following computations. We can easily bound the ratio $\mu_s^k[d_s]/d_s(\eta(s))$ by

$$\frac{\mu_s^k[d_s]}{d_s(\eta(s))} = 1 + \frac{\mu_s^k[d_s] - d_s(\eta(s))}{d_s(\eta(s))} \leq \exp\big(Y_s(\eta(s))\,(1 + c\varepsilon(n))\big), \quad (5.196)$$

where we introduced the random variable

$$Y_s(\sigma) := \frac{1}{|\Lambda_{k,\Theta_{k,s}^-}|} \sum_{i \in \Lambda_k} a_i(s) \left(\mu_s^k\Big[\mathbb{1}_{i \in \Lambda_{k,\Theta_{k,s}^-}}\Big] - \mathbb{1}_{i \in \Lambda_{k,\Theta_{k,s}^-}}(\sigma) \right) \quad (5.197)$$

with $a_i(s) := \exp\big(\beta\,\langle \tilde{h}^i, -e^{\Theta_{k,s}^-} + e^{\Theta_{k,s}^+}\rangle\big) - 1$. Thus, it remains to evaluate the sums in (5.195). For any $u \in \{1, \ldots, t-1\}$, we claim that

$$\sum_{\eta(u),\ldots,\eta(t-1)} \prod_{s=u}^{t-1} \frac{\mu_s^k[d_s]}{d_s(\eta(s))}\, q_s(\eta(s-1), \eta(s))$$

$$\leq \exp\left(\sum_{s=u}^{t-1} \Big(Y_s(\eta(u-1)) + c\varepsilon(n)\tfrac{t-u}{|\Lambda_k|}\Big) \right) \quad (5.198)$$

uniformly for all $\eta(u-1) \in \mathcal{S}[\tilde{x}(u-1)]$. Once we have establish (5.198), it follows that

$$\mathbb{E}_{\mu_0^k}[\psi_t^k] \leq \mu_0^k\Big[\exp\big(\textstyle\sum_{s=0}^{t-1} Y_s\big) \Big]\, \exp\!\Big(c\varepsilon(n)\tfrac{t^2}{|\Lambda_k|}\Big). \quad (5.199)$$

The proof of (5.198) relies one the following observation. For any $u \in \{1, \ldots, t-1\}$ consider an arbitrary chosen but fixed $\sigma \in \mathcal{S}[\tilde{x}(u)]$. Then, for all $j \in \Lambda_{k,\Theta_{k,u-1}^-}(\sigma)$ and arbitrary s,

$$Y_s(\sigma^{j,\Theta_{k,u-1}^+})$$
$$= Y_s(\sigma) + \frac{1}{|\Lambda_{k,\Theta_{k,s}^-}|} \sum_{i \in \Lambda_k} a_i(s)\,\mathbb{1}_{i=j}\left(\mathbb{1}_{\Theta_{k,s}^-=\Theta_{k,u-1}^-} - \mathbb{1}_{\Theta_{k,s}^-=\Theta_{k,u-1}^+}\right)$$
$$\leq Y_s(\sigma) + \frac{c\varepsilon(n)}{|\Lambda_k|}. \quad (5.200)$$

Therefore, we obtain by induction that

$$\sum_{\eta(u),\ldots,\eta(t-1)} \prod_{s=u}^{t-1} \frac{\mu_s^k[d_s]}{d_s(\eta(s))}\, q_s(\eta(s-1), \eta(s))$$

$$\leq \sum_{\sigma \in \mathcal{S}[\tilde{x}(u)]} \exp\!\left(\sum_{s=u}^{t-1} \Big(Y_s(\sigma) + c\varepsilon(n)\tfrac{t-u+1}{|\Lambda_k|}\Big) \right) q_u(\eta(u-1), \sigma)$$

$$\leq \exp\!\left(\sum_{s=u}^{t-1} \Big(Y_s(\eta(u-1)) + c\varepsilon(n)\tfrac{t-u}{|\Lambda_k|}\Big) \right),$$

where we take advantage of (5.200) is the second step.

STEP 3. Finally, we are left with the task to bound $\mu_0^k\left[\exp\left(\sum_{s=0}^{t-1} Y_s\right)\right]$ from above. Let us consider the log-moment generating function $h(u) = \ln \mu_0^k\left[\exp\left(u \sum_{s=0}^{t-1} Y_s\right)\right]$ for $u \geq 0$. Since any function, h, that vanishes at zero, satisfies the bound $h(t) \leq \frac{t^2}{2} \max_{0 \leq s \leq t} h''(s) + t\, h'(0)$, we immediately get that

$$\ln \mu_0^k\left[\exp\left(\sum_{s=0}^{t-1} Y_s\right)\right] \leq \max_{0 \leq u \leq 1} \mathbf{Var}^{t,u}\left[\exp\left(\sum_{s=0}^{t-1} Y_s\right)\right] + \sum_{s=0}^{t-1} \mu_0^k[Y_s], \quad (5.201)$$

where $\mathbf{Var}^{t,u}$ is the variance with respect to the tilted measure $\mu_0^{k,t,u}$ conditioned on $\mathcal{S}[\tilde{x}(0)]$,

$$\mu_0^{k,t,u}(\sigma) = \frac{\mu_0^k(\sigma)\, \exp\left(u \sum_{s=0}^{t-1} Y_s(\sigma)\right)}{\mu_0^k\left[\exp\left(u \sum_{s=0}^{t-1} Y_s\right) \mathbb{1}_{\varrho^k(\sigma)=\tilde{x}(0)}\right]}, \quad \forall \sigma \in \mathcal{S}[\tilde{x}(0)]. \quad (5.202)$$

In Lemma 5.34, we will derive bounds for the expectation, $\mu_0^k[Y_s]$, and the variance, $\mathbf{Var}^{t,u}[Y_s]$, respectively. Notice that the proof of this lemma crucially relies on the fact that the distance of the mesoscopic path x from the boundary of $\times_{k=1}^n \mathcal{M}_{\pi_k}$ is bounded from below by a constant that is independent of N. Consequently, assuming (5.207) we have

$$\sum_{s=0}^{t-1} \mu_0^k[Y_s] \leq c_{21}\,\varepsilon(n) \sum_{s=0}^{t-1} \frac{s}{|\Lambda_k|} = c_{21}\,\varepsilon(n) \frac{t^2}{|\Lambda_k|}.$$

On the other hand, using the Cauchy-Schwartz inequality and assuming (5.208)

$$\mathbf{Var}^{t,u}\left[\exp\left(\sum_{s=0}^{t-1} Y_s\right)\right] \leq \left(\sum_{s=0}^{t-1} \sqrt{\mathbf{Var}^{t,u}[Y_s]}\right)^2 \leq c_{22}\,\varepsilon(n) \frac{t^2}{|\Lambda_k|}.$$

By combining the estimates above, we finally obtain that there exists $c < \infty$ such that

$$\ln \mu_0^k\left[\exp\left(\sum_{s=0}^{t-1} Y_s\right)\right] \leq c\,\varepsilon(n) \frac{t^2}{|\Lambda_k|}. \quad (5.203)$$

This concludes the proof. □

In the remaining part of this subsection we will proof the Lemma 5.34. To start with, notice that the tilted measure, $\mu_0^{k,t,u}$, as well as the canonical measure, $\mu_0^k \equiv \mu_0^{k,0,0}$, can be written as a product measures, $\nu^{k,t,u}$, conditioned on $\mathcal{S}[\tilde{x}(0)]$, i.e.

$$\mu_0^{k,t,u}(\sigma) = \bigotimes_{i \in \Lambda_k} p_i^{k,t,u}\left[\sigma \,|\, \mathcal{S}[\tilde{x}(0)]\right] =: \nu^{k,t,u}\left[\sigma \,|\, \mathcal{S}[\tilde{x}(0)]\right],$$

where for all $r \in \mathcal{S}_0$

$$p_i^{k,t,u}(r) := \frac{\exp\left(\beta\,\langle \tilde{h}^i, e^r\rangle + u \sum_{s=0}^{t-1} |\Lambda_{k,\Theta_{k,s}^-}|^{-1} a_i(s) \mathbb{1}_{r = \Theta_{k,s}^-}\right)}{\sum_{r' \in \mathcal{S}_0} \exp\left(\beta\,\langle \tilde{h}^i, e^{r'}\rangle + u \sum_{s=0}^{t-1} |\Lambda_{k,\Theta_{k,s}^-}|^{-1} a_i(s) \mathbb{1}_{r' = \Theta_{k,s}^-}\right)}.$$

5.4. Lower bounds on capacities

Let us point out that, as a consequence of the choice of the paths $\boldsymbol{x} \in \mathcal{GP}_{A,B}$, there exists $\delta_1 \in (0,1)$ independent of N such that uniformly for all $i \in \Lambda_k$, $t \in \{0, \ldots, \tau_k[b_{\boldsymbol{x}}]\}$ and $u \in [0,1]$ the probabilities $p_i^{k,u}$ satisfy

$$\delta_1 \leq p_i^{k,t,u}(r) \leq 1 - \delta_1, \qquad \forall r \in \mathcal{S}_0. \tag{5.204}$$

A key element in the analysis are sharp large deviation estimates, as well as properties of the involved entropy and corresponding log–moment generating function. Given a subset $\Lambda \subset \Lambda_k$, we define $\varrho_{|\Lambda_k \setminus \Lambda|} : \mathcal{S}_{|\Lambda_k|} \to \mathcal{M}_{(\Lambda_k - \Lambda)/N}$,

$$\sigma \mapsto \varrho_{|\Lambda_k \setminus \Lambda|}(\sigma) := \frac{1}{N} \sum_{i \in \Lambda_k \setminus \Lambda} \delta_{\sigma_i}$$

Note that in later applications we will choose $\Lambda = \{i\}$ or $\Lambda = \{i,j\}$. Further, for any $x \in \mathbb{R}^q$ consider the entropy

$$I_{|\Lambda_k \setminus \Lambda|}^{t,u}(x) := \sup_{s \in \mathbb{R}^q} \left(\langle s, x \rangle - U_{|\Lambda_k \setminus \Lambda|}^{t,u}(s) \right)$$

which is defined as the Legendre-Fenchel transform of the log-moment generating function $U_{|\Lambda_k \setminus \Lambda|}^{t,u} : \mathbb{R}^q \to \mathbb{R}$,

$$t \mapsto U_{|\Lambda_k \setminus \Lambda|}^{t,u}(s) := \frac{1}{N} \ln \nu^{k,t,u} \left[\exp \left(N \langle s, \varrho_{|\Lambda_k \setminus \Lambda|} \rangle \right) \right]$$
$$= \frac{1}{N} \sum_{i \in \Lambda_k \setminus \Lambda} \ln \left(\sum_{r \in \mathcal{S}_0} e^{sr} p_i^{k,t,u}(r) \right).$$

Let us remark that, due to (5.204), a statement analog to Lemma 5.6 can be established. Moreover, by inspecting the proof of [26, Theorem 3.1], it can be shown that the error term in (5.33) is actually of order $1/N$. Hence, for all $x \in \mathrm{ri}(\mathcal{M}_{(\Lambda_k - \Lambda)/N})$,

$$\nu^{k,t,u}\left[\mathbf{1}_{\varrho_{|\Lambda_k \setminus \Lambda|} = x} \right] = \frac{\exp\left(-N I_{|\Lambda_k \setminus \Lambda|}^{t,u}(x)\right)}{\sqrt{(2\pi N)^{q-1} \det\left[\mathrm{Hess}\, U_{|\Lambda_k \setminus \Lambda|}^{t,u}(t^\Lambda(x))\right]_q}} \left(1 + \mathcal{O}\left(\tfrac{1}{N}\right)\right),$$

where $t^\Lambda(x) \in \mathbb{R}^q$ is the solution of $x = \nabla U_{|\Lambda_k \setminus \Lambda|}^{t,u}(t^\Lambda(x))$ such that $\sum_{r \in \mathcal{S}_0} t_r^\Lambda(x) = 0$.

Lemma 5.33. *Suppose $y, y' \in \mathrm{ri}(\mathcal{M}_{(\Lambda_k \setminus \Lambda)/N})$. Then, there exists $c < \infty$, independent of N, such that*

$$\|t^\Lambda(y') - t^\Lambda(y)\| \leq c \|y' - y\|. \tag{5.205}$$

Proof. Due to the fact that $y, y' \in \mathrm{ri}(\mathcal{M}_{(\Lambda_k - \Lambda)/N})$, there exists $\delta \equiv \delta(y, y') > 0$ such that for all $0 \leq s \leq 1$ the minimal component of $s\, y' - (1-s)y$ is at least δ. Recall that in a local coordinate system (5.30) reads

$$\psi(t^*) = \nabla\left(I_{|\Lambda_k \setminus \Lambda|}^{t,u} \circ \psi^{-1}\right)(\psi(x)) \quad \Longleftrightarrow \quad \psi(x) = \nabla\left(U_{|\Lambda_k \setminus \Lambda|}^{t,u} \circ \psi^{-1}\right)(\psi(t^*)).$$

Here, we choose as a local chart $\psi \colon \mathcal{M}_{(\Lambda_k - \Lambda)/N} \to \mathbb{R}^{q-1}$ the linear map $\psi(x) = V^T x$ where the columns of the matrix V consists of the vectors of an orthonormal basis of V_1. Further, we set $\xi := y' - y \in V_1$. Then,

$$\begin{aligned}
&\|\psi(t^\Lambda(y+\xi)) - \psi(t^\Lambda(y))\| \\
&= \left\|\nabla\left(I^{t,u}_{|\Lambda_k \setminus \Lambda|} \circ \psi^{-1}\right)(\psi(y+\xi)) - \nabla\left(I^{t,u}_{|\Lambda_k \setminus \Lambda|} \circ \psi^{-1}\right)(\psi(y))\right\| \\
&\leq \int_0^1 \left\|\operatorname{Hess}\left(I^{t,u}_{|\Lambda_k \setminus \Lambda|} \circ \psi^{-1}\right)(\psi(y) - s\,\psi(\xi))\,\psi(\xi)\right\| ds \\
&\leq \int_0^1 \left\|\left(V^T \operatorname{Hess} U^{t,u}_{|\Lambda_k \setminus \Lambda|}(t^\Lambda(y - s\xi))\, V\right)^{-1} \psi(\xi)\right\| ds \\
&\leq c\,\|\psi(y' - y)\|, \quad\quad\quad\quad\quad\quad\quad\quad\quad\quad\quad\quad (5.206)
\end{aligned}$$

where $c = \delta^{-2} e^{2\beta\varepsilon(n)}$. Note that we used (5.58) in the third step and Lemma 5.6 in the last one. This completes the proof. $\square$

Lemma 5.34. *For all $\underline{x} \in \mathcal{GP}_{A,B}$ and $k = 1, \ldots, n$ there exists constants $c_{21}, c_{22} < \infty$ such that for all $s \in \{0, \ldots, \tau_k[b_{\underline{x}}]\}$*

(i) *the expectation of Y_s with respect to μ_0^k is bounded by*

$$\mu_0^k[Y_s] \leq c_{21}\,\varepsilon(n)\,\frac{s}{|\Lambda_k|}, \quad\quad\quad (5.207)$$

(ii) *the variance of Y_s with respect to $\mu_0^{k,u}$ is bounded by*

$$\mathbf{Var}^u[Y_s] \leq c_{22}\,\frac{\varepsilon(n)}{|\Lambda_k|} \quad\quad\quad (5.208)$$

uniformly for all $u \in [0, 1]$.

Proof. Let us consider an arbitrary but fixed $k \in \{1, \ldots, n\}$ and $s \in \{0, \ldots, \tau_k[b_{\underline{x}}]\}$. Moreover, we set $r = \Theta^-_{k,s}$ and abbreviate $\nu^k \equiv \nu^{k,0,0}$ throughout this proof to lighten notation.

(i) In view of (5.197), it remains to analyze the difference $\left|\mu_s^k[\mathbb{1}_{i\in\Lambda_{k,r}}] - \mu_0^k[\mathbb{1}_{i\in\Lambda_{k,r}}]\right|$ for $s \geq 1$. By a straight forward computation we obtain first of all that, for N large enough,

$$\begin{aligned}
&\frac{\mu_s^k[\mathbb{1}_{i\in\Lambda_{k,r}}] - \mu_0^k[\mathbb{1}_{i\in\Lambda_{k,r}}]}{\mu_s^k[\mathbb{1}_{i\in\Lambda_{k,r}}]} \\
&= \sum_{q\in S_0} \mu_s^k[\mathbb{1}_{i\in\Lambda_{k,q}}]\left(1 - \frac{\nu^k[\mathbb{1}_{\varrho_{|\Lambda_k\setminus\{i\}|} = \tilde{x}(0) - \hat{e}^r}]\,\nu^k[\mathbb{1}_{\varrho_{|\Lambda_k\setminus\{i\}|} = \tilde{x}(s) - \hat{e}^q}]}{\nu^k[\mathbb{1}_{\varrho_{|\Lambda_k\setminus\{i\}|} = \tilde{x}(0) - \hat{e}^q}]\,\nu^k[\mathbb{1}_{\varrho_{|\Lambda_k\setminus\{i\}|} = \tilde{x}(s) - \hat{e}^r}]}\right) \\
&= \sum_{q\in S_0} \mu_s^k[\mathbb{1}_{i\in\Lambda_{k,q}}]\left(1 - \exp(-|\Lambda_k|\,R_1)\left(1 + \mathcal{O}(\tfrac{1}{N})\right)\right) \quad (5.209)
\end{aligned}$$

5.4. Lower bounds on capacities

where the remainder, R_1, is given by

$$R_1 = I^i_{|\Lambda_k|}(\tilde{x}(s) - \hat{e}^r) - I^i_{|\Lambda_k|}(\tilde{x}(s) - \hat{e}^q) - I^i_{|\Lambda_k|}(\tilde{x}(0) - \hat{e}^r) + I^i_{|\Lambda_k|}(\tilde{x}(0) - \hat{e}^q). \tag{5.210}$$

Here, we introduce the notation $I^i_{|\Lambda_k|} \equiv I^{0,0}_{|\Lambda_k|\setminus\{i\}}$. Now, for any $y \in \mathrm{ri}(\mathcal{M}_{(|\Lambda_k|-1)/N})$, let us define $t^i(y) \in \mathbb{R}^q$ as a solution of $y = \nabla U^i_{|\Lambda_k|}(t^i(y))$. As a immediate consequence of (5.29), we have

$$\begin{aligned} I^i_{|\Lambda_k|}(y - \hat{e}^r) - I^i_{|\Lambda_k|}(y - \hat{e}^q) &= \langle t^i(y - \hat{e}^r), \hat{e}^q - \hat{e}^r \rangle + U^i_{|\Lambda_k|}(t^i(y - \hat{e}^q)) \\ &\quad - U^i_{|\Lambda_k|}(t^i(y - \hat{e}^r)) - \langle t^i(y - \hat{e}^r) - t^i(y - \hat{e}^q), y \rangle \\ &= \langle t^i(y - \hat{e}^r), \hat{e}^q - \hat{e}^r \rangle + \mathcal{O}(1/N^2). \end{aligned} \tag{5.211}$$

where we used a Taylor expansion of $U^i_{|\Lambda_k|}(t^i(y - \hat{e}^q))$ and (5.205) in the second step. Therefore,

$$R_1 = \langle t^i(\tilde{x}(s) - \hat{e}^r) - t^i(\tilde{x}(0) - \hat{e}^r), \hat{e}^q - \hat{e}^r \rangle + \mathcal{O}(1/N^2) \leq c \frac{s}{N^2}, \tag{5.212}$$

provided that $c < \infty$ is chosen appropriately. As a result, there exists $c_{21} < \infty$ such that

$$\mu^k_0[Y_s] = \mathcal{O}(\varepsilon(n)) \left| \mu^k_s[\mathbb{1}_{i\in\Lambda_{k,r}}] - \mu^k_0[\mathbb{1}_{i\in\Lambda_{k,r}}] \right| \leq c_{21} \varepsilon(n) \frac{s}{N}. \tag{5.213}$$

(ii) To start with, let us rewrite the variance, $\mathbf{Var}^{t,u}[Y_s]$, as

$$\mathbf{Var}^{t,u}[Y_s] \leq \frac{c\varepsilon(n)}{|\Lambda_{k,r}|} + \frac{1}{|\Lambda_{k,r}|^2} \sum_{\substack{i,j\in\Lambda_k \\ i\neq j}} a_i(s) a_j(s) \mathbf{Cov}^{t,u}[\mathbb{1}_{i\in\Lambda_{k,r}}, \mathbb{1}_{j\in\Lambda_{k,r}}] \tag{5.214}$$

Hence, it suffices to show that the covariance $\mathbf{Cov}^{t,u}[\mathbb{1}_{i\in\Lambda_{k,r}}, \mathbb{1}_{j\in\Lambda_{k,r}}]$ is of order $1/N$ for any distinct $i, j \in \Lambda_k$. By a computation similar to the one in (5.209), we obtain

$$\begin{aligned} &\frac{\mathbf{Cov}^{t,u}[\mathbb{1}_{i\in\Lambda_{k,r}}, \mathbb{1}_{j\in\Lambda_{k,r}}]}{\mu^{k,t,u}_0[\mathbb{1}_{i\in\Lambda_{k,r}} \mathbb{1}_{j\in\Lambda_{k,r}}]} \\ &= \sum_{q\in S_0} \mu^{k,t,u}_0[\mathbb{1}_{i\in\Lambda_{k,q}}] \\ &\quad \times \left(1 - \frac{\nu^{k,u}[\mathbb{1}_{\varrho|\Lambda_k\setminus\{i\}|=\tilde{x}(0)-\hat{e}^r}] \, \nu^{k,u}[\mathbb{1}_{\varrho|\Lambda_k\setminus\{i,j\}|=\tilde{x}(0)-\hat{e}^r-\hat{e}^q}]}{\nu^{k,u}[\mathbb{1}_{\varrho|\Lambda_k\setminus\{i\}|=\tilde{x}(0)-\hat{e}^q}] \, \nu^{k,u}[\mathbb{1}_{\varrho|\Lambda_k\setminus\{i,j\}|=\tilde{x}(0)-2\hat{e}^r}]} \right) \\ &= \sum_{q\in S_0} \mu^{k,t,u}_0[\mathbb{1}_{i\in\Lambda_{k,q}}] \left(1 - \exp(-|\Lambda_k| R_2) \left(1 + \mathcal{O}(\tfrac{1}{N})\right) \right). \end{aligned} \tag{5.215}$$

By introducing the notation $I_{|\Lambda_k|}^{i,j} \equiv I_{|\Lambda_k \setminus \{i,j\}|}^{t,u}$, the remainder can be written as

$$\begin{aligned} R_2 &= I_{|\Lambda_k|}^{i,j}(\tilde{x}(0) - 2\,\hat{e}^r) - I_{|\Lambda_k|}^{i,j}(\tilde{x}(0) - \hat{e}^r - \hat{e}^q) \\ &\quad - I_{|\Lambda_k|}^{i}(\tilde{x}(0) - \hat{e}^r) + I_{|\Lambda_k|}^{i}(\tilde{x}(0) - \hat{e}^q) \\ &= \langle t^{i,j}(\tilde{x}(0) - 2\,\hat{e}^r) - t^{i}(\tilde{x}(0) - \hat{e}^r), \hat{e}^q - \hat{e}^r\rangle + \mathcal{O}(1/N^2) \\ &= \mathcal{O}(1/N^2). \end{aligned} \tag{5.216}$$

In the last step, we took advantage of (5.205) and the fact that $t^{i}(\tilde{x}(0) - \hat{e}^r) \in \partial I^{i,j}(y)$ with $\|\tilde{x}(0) - \hat{e}^r - y\| \le \sqrt{q}/N$. Thus, by combining the estimates above, the assertion of (5.208) follows. □

5.5. Sharp estimates on metastable exit times and exponential distribution

The representation (1.49) of the averaged mean hitting time, $\mathbb{E}_{\nu_{A,B}}[\tau_B]$, in terms of the capacity, $\mathrm{cap}(A, B)$, and the equilibrium potential, $h_{A,B}$, has the advantage that it suffices to compute precisely only $\mathrm{cap}(A, B)$ and to establish some rough bounds on $h_{A,B}$. In the previous sections we have established upper and lower bounds on $\mathrm{cap}(A, B)$ that coincides in the limit when N tends to infinity. Hence, our main objective now is to show that the equilibrium potential is close to one in the neighborhood, $\mathcal{U}_\delta(A)$, of the starting set A, whereas outside this neighborhood it is sufficiently small to ensure that the summation of $\mu_N(\sigma) h_{A,B}(\sigma)$ over all $\sigma \in \mathcal{S}_N \setminus \mathcal{U}_\delta(A)$ compare to $\mu_N[A]$ is negligible. Notice that in general such a behavior of the equilibrium potential, $h_{A,B}$, depends crucially on the choice of the subsets A and B.

To start with, let $m, m' \in \Gamma^n$ be two local minima of the mesoscopic free energy F^n. Further, let $z \in \Gamma^n$ be the lowest saddle point of index one between m from m'. As a consequence of Lemma 5.8, for any n, every mesoscopic critical point m determines uniquely a corresponding macroscopic critical point $m \equiv m(m) \in \Gamma_N$ and vice versa. Recall that the value of the mesoscopic free energy and macroscopic free energy coincides at critical points, i.e. $F_N(m) = F^n(m)$. Let us consider the sets $A \equiv A^n = \mathcal{S}^n[m] \subset \mathcal{S}_N$ and $B \equiv B^n = \mathcal{S}^n[m'] \subset \mathcal{S}_N$. In view of (5.67) and (5.124), there exists $\mathfrak{C}_1 > 0$ such that for all $n \in \mathbb{N}$ there exists $N(n) \in \mathbb{N}$ and some $a_n < \infty$ such that

$$\mathbb{P}_{\mu_A}[\tau_B < \tau_A] \le a_n\, N^{n(q-1)/2-1} \exp\!\left(-\beta N\left(F_N(z) - F_N(m)\right)\right) \le \mathrm{e}^{-\beta\mathfrak{C}_1 N}, \tag{5.217}$$

for all $N \ge N(n)$. By choosing $\mathfrak{C}_1$ appropriately, we can ensure that the last inequality holds true uniformly for all minima m, m' of F^n. For this reason, for every n, the set $\mathcal{M}^n := \{\mathcal{S}^n[m] \mid m \in \Gamma^n \text{ local minima of } F^n\}$ is a suitable candidate for the set of

5.5. Sharp estimates on metastable exit times and exponential distribution

metastable sets. In the following lemma we show that, if n is chosen large enough, $\mathcal{M}^n$ satisfies (3.1) with $\mathfrak{C} := \mathfrak{C}_1 - 2\beta\varepsilon(n)$.

Lemma 5.35. *For all $n \in \mathbb{N}$ there exists $N(n) \in \mathbb{N}$ such that for all $X = \mathcal{S}^n[\boldsymbol{x}] \notin \mathcal{M}^n$ with $\boldsymbol{x} \in \Gamma^n$*

$$\mathbb{P}_{\mu_X}\left[\tau_{\mathcal{M}^n} < \tau_X\right] \geq e^{-2\beta\varepsilon(n)N}, \qquad \forall N \geq N(n). \tag{5.218}$$

Proof. For every $\boldsymbol{x} \in \Gamma^n$ there exists a local minimum m of F^n and a self-avoiding mesoscopic path $\underline{\boldsymbol{x}}$ from $\boldsymbol{x}$ to m such that the mesoscopic free energy is non-increasing along the path. By exploiting the fact that the Dirichlet form is monotone in the transition probabilities provides an immediate lower bound for $\mathbb{P}_{\mu_X}\left[\tau_{\mathcal{M}^n} < \tau_X\right]$ in terms of the capacity $\text{cap}^{\boldsymbol{x}}(M,X)$ of a chain where all $p_N(\sigma,\eta)$ are set to zero whenever $\sigma, \eta \notin \mathcal{S}^n[\underline{\boldsymbol{x}}]$, i.e.

$$\text{cap}(X, \mathcal{M}^n) \geq \inf_{f \in \mathcal{H}_{X,M}} \frac{1}{2} \sum_{\sigma, \eta \in \mathcal{S}^n[\underline{\boldsymbol{x}}]} \mu_N(\sigma)\, p_N(\sigma,\eta)\, (f(\sigma)-f(\eta))^2 = \text{cap}^{\boldsymbol{x}}(X, A).$$

Here $A = \mathcal{S}^n[m]$ and $\mathcal{S}^n[\underline{\boldsymbol{x}}]$ denotes the set of configurations that correspond to the mesoscopic path. On the other hand, the Berman-Konsowa principle implies

$$\text{cap}^{\boldsymbol{x}}(X, A) \geq \text{CAP}^{n,\boldsymbol{x}}(\boldsymbol{x}, m) \left(\sum_{t=0}^{|\underline{\boldsymbol{x}}|-1} \mathbb{E}^{f^{\boldsymbol{x}}}[\Phi_t^{\boldsymbol{x}}] \left(g_{\boldsymbol{x},m}(\boldsymbol{x}(t)) - g_{\boldsymbol{x},m}(\boldsymbol{x}(t+1)) \right) \right)^{-1}$$

where $\Phi_t^{\boldsymbol{x}}$ is defined in (5.135) and $g_{\boldsymbol{x},m}$ is the mesoscopic harmonic function along the one-dimensional path.

For the subordinate unit flow $f^{\boldsymbol{x}}$ we choose the flow that is induced by a simple directed random walk on $\mathcal{S}^n[\underline{\boldsymbol{x}}]$, i.e. for $k = 1, \ldots, n$ such that $\|x^k(t+1) - x^k(t)\| = 2/N$ and $r = \Theta_{k,t}^-$

$$f^{\boldsymbol{x}}(\sigma, \eta) = \frac{\mathbb{1}_{\eta \in \Lambda_{k,r}(\sigma)}}{|\mathcal{S}^n[\boldsymbol{x}(t)]|\, |\Lambda_{k,r}(\sigma)|} \qquad \forall \sigma \in \mathcal{S}^n[\boldsymbol{x}(t)],\, \eta \in \mathcal{S}^n[\boldsymbol{x}(t+1)].$$

Since $\mathcal{Q}^n(\boldsymbol{x})/\mu_N(\sigma) \leq e^{\beta\varepsilon(n)N}|\mathcal{S}^n[\boldsymbol{x}]|$ for all $\sigma \in \mathcal{S}^n[\boldsymbol{x}]$ and $\boldsymbol{x} \in \Gamma^n$, our choice of $f^{\boldsymbol{x}}$ gives rise to $\Phi_t^{\boldsymbol{x}}(\sigma) \leq e^{\beta\varepsilon(n)N}$ uniformly for all $\sigma \in \mathcal{S}^n[\boldsymbol{x}(t)]$ and $t = 0, \ldots, |\underline{\boldsymbol{x}}|-1$. Using that the mesoscopic capacity along a one-dimensional path can be computed explicitly, see e.g. [11]

$$\text{CAP}^{n,\boldsymbol{x}}(\boldsymbol{x}, m) = \left(\sum_{t=0}^{|\underline{\boldsymbol{x}}|-1} \frac{1}{\mathcal{Q}^n(\boldsymbol{x}(t))\, r^n(\boldsymbol{x}(t), \boldsymbol{x}(t+1))} \right)^{-1} \geq c\, \frac{\mathcal{Q}^n(\boldsymbol{x})}{N\sqrt{N}}$$

where we used that $r^n(\boldsymbol{x}(t), \boldsymbol{x}(t+1)) \geq c'/N$ and $F^n(\boldsymbol{x}(t)) - F^n(m) \geq c''(|\underline{\boldsymbol{x}}| - t)^2/N^2$ in a neighborhood of a local minimum m. Hence, there exists $N(n) \in \mathbb{N}$ such that for all $N \geq N(n)$

$$\mathbb{P}_{\mu_X}\left[\tau_{\mathcal{M}^n} < \tau_X\right] \geq \frac{c}{N\sqrt{N}} e^{-\beta\varepsilon(n)N} \geq e^{-2\beta\varepsilon(n)N}.$$

This concludes the proof. □

Proof of Theorem 5.5. (i) For δ specified in Theorem 5.5, we define

$$\mathcal{U}_\delta := \{x \in \Gamma^n \mid F^n(x) < F^n(m) + \delta\} = \mathcal{U}_\delta(m) \cup \bigcup_{m' \in M} \mathcal{U}_\delta(m') \qquad (5.219)$$

where $\mathcal{U}_\delta(m)$ denotes the connected component of $\mathcal{U}_\delta$ containing m. Inspecting the proof of Theorem 3.10, we obtain that

$$\mathbb{E}_{\nu_{A,B}}[\tau_B] = \frac{1}{\mathrm{cap}(A,B)} \sum_{x \in \mathcal{U}_\delta(m)} \mathcal{Q}^n(x) \left(1 + o_N(1)\right). \qquad (5.220)$$

Note that, in view of (5.218), we only have to choose $c_1 > 1 + \alpha^{-1}$ for $\alpha^{-1} = \mathcal{O}(q)$ such that $c_2 > 2\beta\varepsilon(n)$. It remains to evaluate the right-hand side of (5.220). Analog to the proof of Proposition 5.21, a standard approximation of the sum by an integral yields

$$\begin{aligned}
\sum_{x \in \mathcal{U}_\delta(m)} \mathcal{Q}^n(x) &= \mathcal{Q}^n(m) \sum_{x \in \mathcal{U}_\delta(m)} \exp\left(-\tfrac{\beta N}{2} \langle x, A^n(m) x \rangle\right) \\
&= \mathcal{Q}^n(m) \left(\frac{N^{q-1}}{\sqrt{q}}\right)^n \int_{V_1} \exp\left(-\tfrac{\beta N}{2} \langle x, A^n(m) x \rangle\right) dx \, (1 + o_N(1)) \\
&= \mathcal{Q}^n(m) \left(\frac{(2\pi N)^{q-1}}{q \beta^{q-1}}\right)^{n/2} \left(\sqrt{\det\left(A^n(m)\right)}\right)^{-1/2} (1 + o_N(1))
\end{aligned} \qquad (5.221)$$

Thanks to (5.67) and (5.68), we get that

$$\sum_{x \in \mathcal{U}_\delta(m)} \mathcal{Q}^n(x) = \frac{\exp\left(-\beta N F_N(m)\right) (1 + o_N(1))}{Z_N \sqrt{\det\left(I_q - 2\beta \left(D(z) - \mathbb{E}_h[u^1(2\beta z) \cdot u^1(2\beta z)^T]\right)\right)}}. \qquad (5.222)$$

Thus, by combining (5.220) with (5.221) and the expression for the capacity that is given in (5.20) concludes the proof of (i).

Moreover, (ii) and (iii) follows easily from Corollary 3.12 and Theorem 3.13. □

Bibliography

1. J.M.G. Amaro de Matos, A.E. Patrick, and V.A. Zagrebnov, *Random infinite-volume Gibbs states for the Curie-Weiss random field Ising model*, J. Stat. Phys. **66** (1992), no. 1-2, 139–164.
2. S. Arrhenius, *Über die Reaktionsgeschwindigkeit bei der Inversion von Rohrzucker durch Säuren*, Z. Phys. Chem. **4** (1889), 226–248.
3. G. Ben Arous and R. Cerf, *Metastability of the tree dimensional Ising model on a torus at very low temperatures*, Electron. J. Probab. **1** (1996), no. 10, 1–55.
4. K.A. Berman and M.H. Konsowa, *Random-paths and cuts, electrical networks, and reversible Markov-chains*, SIAM J. Discrete Math. **3** (1990), no. 3, 311–319.
5. A. Bianchi, A. Bovier, and D. Ioffe, *Pointwise estimates and exponential laws in metastable systems via coupling methods*, SFB611 Preprint **461** (2009), 1–28.
6. _____, *Sharp asymptotics for metastability in the random field Curie-Weiss model*, Elect. J. Probab. **14** (2009), 1541–1603.
7. A. Bianchi and A. Gaudillière, *Metastable states, quasi-stationary and soft measures, mixing time asymptotics via variational principles*, arXiv:1103.1143 (2011), 1–32.
8. R.M. Blumenthal and R.K. Getoor, *Markov Processes and Potential Theory*, 1 ed., Pure and Applied Mathematics, vol. 29, Academic Press, 1968.
9. A. Bovier, *Metastability and ageing in stochastic dynamics*, Dynamics and Randomness II, Nonlinear Phenomena and Complex Systems, vol. 10, Kluwer Academic Publisher, 2004, pp. 17–80.
10. _____, *Metastability: A potential theoretic approach*, Proceedings of the International Congress of Mathematicians, vol. 3, European Mathematical Society Publishing House, 2006, pp. 499–518.

11. _____, *Metastability*, Methods of Contemporary Mathematical Statistical Physics, Lecture Notes in Mathematics, vol. 1970, Springer-Verlag, 2009, pp. 177–221.

12. _____, *Metastability: From mean field models to SPDEs*, to appear in Probability in Complex Physical Systems, Springer Proceedings in Mathematics, vol. 11, Springer-Verlag, 2011, pp. 1–16.

13. A. Bovier, F. den Hollander, and F.R. Nardi, *Sharp asymptotics for Kawasaki dynamics on a finite box with open boundary*, Probab. Theory Rel. Fields **135** (2006), no. 2, 265–310.

14. A. Bovier, F. den Hollander, and C. Spitoni, *Homogenous nucleation for Glauber and Kawasaki dynamics in large volumens at low temperatures*, Ann. Probab. **38** (2010), no. 2, 661–713.

15. A. Bovier, M. Eckhoff, V. Gayrard, and M. Klein, *Metastability in stochastic dynamics of disordered mean-field models*, Probab. Theory Rel. Fields **119** (2001), no. 1, 99–161.

16. _____, *Metastability and low lying spectra in reversible Markov chains*, Commun. Math. Phys. **228** (2002), no. 2, 219–255.

17. _____, *Metastability in reversible diffusion processes. I. Sharp asymptotics for capacities and exit times*, J. Europ. Math. Soc. **6** (2004), no. 4, 399–424.

18. A. Bovier, V. Gayrard, and M. Klein, *Metastability in reversible diffusion processes. II. Precise asymptotics for small eigenvalues*, J. Europ. Math. Soc. **7** (2005), no. 1, 69–99.

19. A. Bovier and F. Manzo, *Metastability in Glauber dynamics in the low-temperature limit: Beyond exponential asymptotics*, J. Stat. Phys. **107** (2002), no. 3/4, 757–779.

20. S. Brassesco, E. Olivieri, and Vares M.E., *Couplings and asymptotic exponentiality of exit times*, J. Stat. Phys **93** (1998), no. 1/2, 393–404.

21. C.J. Burke and M. Rosenblatt, *A Markovian function of a Markov chain*, Ann. Math. Statist. **29** (1958), no. 4, 1112–1122.

22. D. Capocaccia, M. Cassandro, and E. Olivieri, *A study of metastability in the Ising model*, Commun. Math. Phys **39** (1974), 185–205.

23. M. Cassandro, A. Galves, E. Olivieri, and M.E. Vares, *Metastable behaviour of stochastic dynamics - A pathwise approach*, J. Stat. Phys. **35** (1984), no. 5-6, 603–634.

24. M. Cassandro, E. Olivieri, and P. Picco, *Small random perturbations of infinite dimensional dynamical systems and nucleation theory*, Ann. Inst. H, Poincaré **44** (1986), no. 4, 343–396.

25. R. Cerf and F. Manzo, *Nucleation and growth for the Ising model in d dimensions at very low temperatures*, arXiv:1102.1741 (2011), 1–84.

26. N.R. Chaganty and J. Sethuraman, *Multidimensional large deviation local limit theorems*, J. Multivariate Anal. **20** (1986), no. 2, 190–204.

27. K.L. Chung, *Green, Brown, and Probability & Brownian Motion on the Line*, 1 ed., World Scientific Publishing, 2002.

28. E.N.M. Cirillo, F.R. Nardi, and C. Spitoni, *Metastability for reversible probabilistic cellular automata with self-interaction*, J. Stat. Phys. **132** (2008), no. 3, 431–471.

29. E.B. Davies, *Metastability and the Ising model*, J. Stat. Phys. **27** (1982), no. 4, 657–675.

30. _____, *Metastable states of symmetric Markov semigroups I*, Proc. London Math. Soc. **45** (1982), no. 3, 133–150.
31. _____, *Metastable states of symmetric Markov semigroups II*, J. London Math. Soc. **26** (1982), no. 2, 541–556.
32. P. Dehghanpour and R.H. Schonmann, *Metropolis dynamics relaxation via nucleation and growth*, Commun. Math. Phys. **188** (1997), 89–119.
33. _____, *A nucleation-and-growth model*, Probab. Theory Relat. Field **107** (1997), 123–135.
34. F. den Hollander, F.R. Nardi, E. Olivieri, and E. Scoppola, *Droplet growth for three-dimensional Kawasaki dynamics*, Probab. Theory Relat. Fields **125** (2003), 153–194.
35. F. den Hollander, E. Olivieri, and E. Scoppola, *Metastability and nucleation for conservative dynamics*, J. Math. Phys. **41** (2000), no. 3, 1424–1498.
36. P. Deuflhard, W. Huisinga, A. Fischer, and Ch. Schütte, *Identification of almost invariant aggregates in reversible nearly uncoupled Markov chains*, Linear Algebra Appl. **315** (2000), 39–59.
37. P. Deuflhard and M. Weber, *Robust perron cluster analysis in conformation dynamics*, Linear Algebra Appl. **398** (2005), 161–184.
38. P.G.L. Dirichlet, *Über einen neuen Ausdruck zur Bestimmung der Dichtigkeit einer unendlich dünnen Kugelschale, wenn der Werth des Potentials derselben in jedem Punkt ihrer Oberfläche gegeben ist*, Königl. Akad. d. Wiss. **1850** (1852), 99–116.
39. R.L. Dobrushin and S.B. Shlosman, *Large and moderate deviations in the Ising model*, Adv. Soviet Math. **20** (1994), 91–219.
40. D. Donsker and S.R.S Varadhan, *On the principal eigenvalue of second-order elliptic differential operators*, Comm. Pure Appl. Math. **29** (1976), 143–157.
41. J.L. Doob, *Semimartingales and subharmonic functions*, Trans. Amer. Math. Soc. **77** (1954), no. 1, 86–121.
42. _____, *Discrete potential theory and boundaries*, J. Math. Mech. **8** (1959), no. 3, 433–458; erratum 993.
43. _____, *Classical Potential Theory and Its Probabilistic Counterpart*, 1 ed., Grundlehren der mathematischen Wissenschaften, vol. 262, Springer-Verlag, 1984.
44. P.G. Doyle and J.L. Snell, *Random Walks and Electric Networks*, 1 ed., Carus Mathematical Monographs, vol. 22, Mathematical Association of America, 1984.
45. M. Eckhoff, *Precise asymptotics of small eigenvalues of reversible diffusions in the metastable regime*, Ann. Probab. **33** (2005), no. 1, 244–299.
46. R. S. Ellis, *Entropy, Large Deviations, and Statistical Mechanics*, Grundlehren der mathematischen Wissenschaften, vol. 271, Springer, Berlin, 1985.
47. H. Eyring, *The activated complex in chemical reations*, J. Chem. Phys. **3** (1935), 107–115.
48. Martinelli F., Olivieri E., and Scoppola E., *Small random perturbations of finite- and infinite-dimensional dynamical systems: Unpredictability of exit times*, J. Stat. Phys. **55** (1989), no. 4/5, 477–504.

49. W. Feller, *An Introduction to Probability Theory and Its Application*, 2 ed., vol. 2, John Willey & Sons, 1970.
50. L.R. Fontes, P. Mathieu, and P. Picco, *On the averaged dynamics of the random field Curie-Weiss model*, Ann. Appl. Probab. **10** (2000), no. 4, 1212–1245.
51. M.I. Freidlin and A.D. Wentzell, *Random Perturbation of Dynamical Systems*, 2 ed., Grundlehren der mathematischen Wissenschaften, vol. 260, Springer, 1998.
52. A. Galves, E. Olivieri, and M.E. Vares, *Metastability for a class of dynamical systems subject to small random perturbations*, Ann. Probab. **15** (1987), no. 4, 1288–1305.
53. A. Gaudillière, *Conderse physics applied to Markov chains – A brief introduction to potential theory*, arXiv:0901.3053 (2008), 1–49.
54. A. Gaudillière, F. den Hollander, F.R. Nardi, E. Olivieri, and E. Scoppola, *Ideal gas approximation for a two-dimensional rarefied gas under Kawasaki dynamics*, Stoch. Process. Appl. **119** (2009), 737–774.
55. B. Gaveau and L.S. Schulman, *Theory of nonequilibrium first-order phase transitions for stochastic dynamics*, J. Math. Phys. **39** (1998), no. 3, 1517–1533.
56. B. Gentz and N. Berglund, *Anomalous behavior of the Kramers rate at bifurcations in classical fild theory*, J. Phys. A **42** (2009), no. 5, 052001:1–9.
57. H. Grabert, *Projection Operator Techniques in Nonequilibrium Systems*, Springer Tracts in Modern Physics, vol. 95, Springer-Verlag, 1982.
58. H. Grabert, P. Hänggi, and P. Talkner, *Microdynamics and nonlinear stochastic processes of gross variables*, J. Stat. Phys. **22** (1980), no. 5, 537–552.
59. G. Green, *An Essay on the Application of Mathematical Analysis to the Theory of Electricity and Magnetism*, 1 ed., Nottingham, 1828.
60. R. Grone, K.H. Hoffmann, and P. Salamon, *An interlacing theorem for reversible Markov chains*, J. Phys. A: Math. Theor **41** (2008), no. 212002, 1–7.
61. P. Hänggi, P. Talkner, and Borkovec M., *Reaction-rate theory: fifty years after Kramers*, Re. Mod. Phys. **62** (1990), no. 2, 251–341.
62. B. Helffer, M. Klein, and F. Nier, *Quantitative analysis of metastability in reversible diffusion processes via a Witten complex approach*, Mat. Contemp. **26** (2004), 1–45.
63. B. Helffer and F. Nier, *Hypoelliptic Estimates and Spectral Theory for Fokker-Planck Operators and Witten Laplacians*, 1 ed., Lecture Notes in Mathematics, vol. 1862, Springer-Verlag, 2005.
64. Lieberstein H.M., *Theory of Partial Differential Equations*, 1 ed., Mathematics in Science and Engineering, vol. 93, Academic Press, 1972.
65. R.A. Holley, S. Kusuoka, and S.W. Stroock, *Asymptotics of the spectral gap with applications to the theory of simulated annealing*, J. Funct. Anal. **83** (1989), 333–347.
66. W. Huisinga, S. Meyn, and Ch. Schütte, *Phase transitions and metastability in Markovian and molecular systems*, Ann. Appl. Probab. **14** (2004), no. 1, 419–458.
67. G.A. Hunt, *Markoff processes and potentials. I*, Illinois J. Math. **1** (1957), no. 1, 44–93.

68. ———, *Markoff processes and potentials. II*, Illinois J. Math. **1** (1957), no. 3, 316–369.
69. ———, *Markoff processes and potentials. III*, Illinois J. Math. **2** (1958), no. 2, 151–213.
70. G. Iacobelli and C. Kuelske, *Metastates in finite-type mean-field models: Visibility, invisibility, and random restoration of symmetry*, J. Stat. Phys. **140** (2010), no. 1, 27–55.
71. J.D. Jackson, *Classical Electrodynamics*, 3 ed., Wiley, 1999.
72. S. Kakutani, *Two-dimensional Brownian motion and harmonic functions*, Proc. Imp. Acad. Tokyo **20** (1944), no. 10, 706–714.
73. I. Karatzas and S.E. Shreve, *Brownian Motion and Stochastic Calculus*, 2 ed., Graduate texts in mathematics, vol. 113, Springer-Verlag, 1991.
74. T. Kato, *On the upper and lower bounds of eigenvalues*, J. Phys. Soc. Japan **4** (1949), 334–339.
75. O.D. Kellogg, *Foundations of Potential Theory*, 1 ed., Grundlehren der mathematische Wissenschaften, vol. 31, Springer-Verlag, 1929.
76. J.G. Kemeny, J.L. Snell, and A.W. Knapp, *Denumerable Markov Chains*, 2 ed., Graduate texts in Mathematics, vol. 40, Springer-Verlag, 1976.
77. Y. Kifer, *A discrete-time version of the Wentzell-Friedlin theory*, Ann. Probab. **18** (1990), no. 4, 1676–1692.
78. H.A. Kramers, *Brownian motion in a field of force and the diffusion model of chemical reactions*, Physica **7** (1940), no. 4, 284–304.
79. C. Kulske, *Metastates in disordered mean-field models: Random field and Hopfield models*, J. Stat. Phys. **88** (1997), no. 5-6, 1257–1293.
80. H. Lebesgue, *Conditions de régularité, conditions d'irrégularité, conditions d'impossibilité dans le problème de Dirichlet*, Comp. Rendue Acad. Sci. **178** (1924), no. 1, 349–254.
81. D.A. Levin, M.J. Luczak, and Y. Peres, *Glauber dynamics for the mean-field Ising model: Cutoff, critical power law, and metastability*, Probab. Theory Rel. Fields **146** (2010), no. 1-2, 223–265.
82. D.A. Levin, Y. Peres, and E.L. Wilmer, *Markov Chain and Mixing Times*, 1 ed., American Mathematical Society, 2008.
83. R. Lyons and Y. Peres, *Probability on Trees and Networks*, Cambridge University Press, 2011, In preparation. Current version available at http://mypage.iu.edu/~rdlyons/.
84. P. Mathieu, *Spectra, exit times and long time asymptotics in the zero white noise limit*, Stoch. and Stoch. Rep. **55** (1995), 1–20.
85. P. Mathieu and P. Picco, *Metastability and convergence to equilibrium for the random field Curie-Weiss model*, J. Stat. Phys. **91** (1998), no. 3/4, 679–732.
86. L. Miclo, *Comportement de spectres d'opérateurs de schrödinger à basse température*, Bull. Sci. Math. **119** (1995), 529–553.
87. L. Mittag and M.J. Stephen, *Mean-field theory of the many component Potts model*, J. Phys. A: Math. Nucl. Gen. **7** (1974), no. 9, L109–L112.
88. C.St.J.A. Nash-Williams, *Random walk and electric currents in networks*, Proc. Cambridge Philos. Soc. **55** (1959), no. 2, 181–194.

89. E.J. Neves and R.H. Schonmann, *Critical droplets and metastability for a Glauber dynamics at very low temperature*, Commun. Math. Phys. **137** (1991), 209–230.
90. _____, *Behavior of droplets for a class of Glauber dynamics at very low temperatur*, Probab. Theory Relat. Fields **91** (1992), 331–354.
91. J.R. Noris, *Markov Chains*, 1 ed., Cambridge University Press, 1997.
92. E. Olivieri and E. Scoppola, *Markov chains with exponentially small transition probabilities: First exit problem from a general domain. I. The reversible case*, J. Stat. Phys. **79** (1995), no. 3/4, 613–647.
93. _____, *Markov chains with exponentially small transition probabilities: First exit problem from a general domain. II. The general case*, J. Stat. Phys **84** (1996), no. 5/6, 987–1041.
94. E. Olivieri and M.E. Vares, *Large Deviations and Metastability*, 1 ed., Encyclopedia of Mathematics and its Applications, vol. 100, Cambridge University Press, 2005.
95. A.M. Ostrowski, *A quantitative formulation of Sylvesters's law of inertia*, Proc. Natl. Acad. Sci. U.S.A. **45** (1959), no. 5, 740–744.
96. O. Penrose and J.L. Lebowitz, *Rigorous treatment of metastable states in the van der Waals-Maxwell theory*, J. Stat. Phys. **3** (1971), no. 2, 211–236.
97. _____, *Towards a rigorous molecular theory of metastability*, Fluctuation Phenomenon, Studies in Statistical Mechanics, vol. 7, North-Holland Publishing Company, 1979, pp. 293–340.
98. H. Poincaré, *Théorie du Potentiel Newtonien*, 1 ed., Carré & Naud, 1899.
99. E. Pollak and P. Talkner, *Reaction rate theory: What it was, where is it today, and where is it going?*, Chaos **15** (2005), no. 026116, 1–11.
100. G. Pólya, *Über eine Aufgabe der Wahrscheinlichkeitsrechnung betreffend die Irrfahrt im Straßennetz*, Math. Ann. **84** (1921), no. 1-2, 149–160.
101. S.C. Port and C.J. Stone, *Brownian Motion and Classical Potential Theory*, 1 ed., vol. 29, Academic Press, 1978.
102. N. Privault, *Potential Theory in Classical Probability*, Quantum Potential Theory, Lecture Notes in Mathematics, vol. 1954, Springer-Verlag, 2008, pp. 3–59.
103. R. T. Rockafellar, *Convex Analysis*, Princeton Mathematical Series, vol. 28, Princeton University Press, Princeton, NJ, 1972.
104. Y. Saad, *Numerical Methods for Large Eigenvalue Problems*, 1 ed., Classics in Applied Mathematics, vol. 66, SIAM, 2011.
105. R.H. Schonmann and S.B. Shlosmann, *Wulff droplets and the metastable relaxation of kinetic Ising models*, Commun. Math. Phys. **194** (1998), 389–462.
106. J. Schwinger, DeRaad L.L.Jr., K.A. Milton, and W. Tsai, *Classical Electrodynamics*, 1 ed., Perseus Books Group, 1998.
107. M. Sugiura, *Metastable behaviors of diffusion processes with small parameter*, J. Math. Soc. Japan **47** (1995), no. 4, 755–788.
108. A. Telcs, *The Art of Random Walks*, 1 ed., Lecture Notes in Mathematics, vol. 1885, Springer-Verlag, 2006.

109. G. Temple, *The accuracy of Rayleigh's method of calculating the natural frequencies of vibrating systems*, Proc. Roy. Soc. London Ser. A **211** (1958), 204–224.
110. H. Thorisson, *Coupling, stationarity, and regeneration*, 1 ed., Probability and its Applications, Springer-Verlag, 2000.
111. A.D. Ventcel and M.I. Freidlin, *On small random perturbations of dynamical systems*, Russ. Math. Surveys **25** (1970), 1–55.
112. A.D. Venttsel', *On the asymptotic of the greatest eigenvalue of a second-order elliptic differential operator with a small parameter in the higher derivatives*, Soviet Math. Dokl. **13** (1972), 13–18.
113. _____, *Formulae for eigenfunctions and eigenmeasures associated with a Markov process*, Theory Probab. Appl. **18** (1973), 1–26.
114. W. Woess, *Random Walks on Infinite Graphs and Groups*, 1 ed., Cambridge Tracts in Mathematics, vol. 138, Cambridge University Press, 2000.
115. F.Y. Wu, *The Potts model*, Rev. Mod. Phys. **54** (1982), no. 1, 235–268.
116. R. Zwanzig, *Nonlinear generalized Langevin equations*, J. Stat. Phys. **9** (1973), no. 3, 215–220.

i want morebooks!

Buy your books fast and straightforward online - at one of world's fastest growing online book stores! Environmentally sound due to Print-on-Demand technologies.

Buy your books online at
www.get-morebooks.com

Kaufen Sie Ihre Bücher schnell und unkompliziert online – auf einer der am schnellsten wachsenden Buchhandelsplattformen weltweit! Dank Print-On-Demand umwelt- und ressourcenschonend produziert.

Bücher schneller online kaufen
www.morebooks.de

VDM Verlagsservicegesellschaft mbH
Heinrich-Böcking-Str. 6-8   Telefon: +49 681 3720 174   info@vdm-vsg.de
D - 66121 Saarbrücken   Telefax: +49 681 3720 1749   www.vdm-vsg.de

Printed by Books on Demand GmbH, Norderstedt / Germany